GREEN LIBERALISM

Green liberalism

The free and the green society

Marcel Wissenburg

Routledge
Taylor & Francis Group

LONDON AND NEW YORK

First published in 1998 by UCL Press

Reprinted 2003 by Routledge
11 New Fetter Lane
London, EC4P 4EE

*Routledge is an imprint of the
Taylor & Francis Group*

British Library Cataloguing in Publication Data
A CIP record for this book is available from the British Library.

Library of Congress Cataloging-in-Publication data are available.

ISBNs:
1-85728-848-3 HB
1-85728-849-1 PB

Typeset in Bembo by Graphicraft Typesetters Ltd, Hong Kong.
Printed and bound by Antony Rowe Ltd, Eastbourne

for Grahame Lock

Contents

CONTENTS

Acknowledgements

This book is one of the tangible results of a research project called 'the free and the green society', which I carried out under the supervision of Professor G.E. Lock. The original project proposal came in first in the 1994 open competition of REOB, the Foundation for Law and Public Policy, which is part of NWO, The Netherlands Organization for Scientific Research. Consequently, the project (from then on known as 415-31-001) was subsidized for the three-year period of May 1995 to May 1998 by REOB, and I became employed for that same period by NWO. Apart from the obligatory but nonetheless heartfelt words of (deep) appreciation for the referees who emphatically supported my proposal, even though they had read it, for the Board of REOB, who for reasons unclear to me followed the referees' advice, and for NWO as an institution, I would like to take this opportunity to express my sincerest thanks to all those individuals at REOB who answered my most obtuse questions, handled my requests for travel grants, actually read my reports on these trips, and in general did everything to support my work and absolutely nothing to hinder it. I also explicitly want to thank the anonymous employees of NWO's financial administration for just doing their job, i.e. paying my salary on time. I have known less efficient organizations - and few that were more efficient.

Most of the chapters in this book evolved out of papers presented at conferences. After presentation, several of these papers were also transformed into and published as articles; the comments I received from anonymous referees were often used to improve both the articles and this book. Chapter 1, Sections 1.4 to 1.7 and Chapter 2, Section 2.6, are permutations of Wissenburg (1996). Chapter 1, Section 1.6 discusses ideas dealt with at (more) length in Van Hees and Wissenburg (forthcoming). Chapter 2, Sections 2.1 to 2.5 originate in Wissenburg (1997a). Chapter 3, Sections 3.2 to 3.4 combine two texts: Wissenburg (1997b) and Wissenburg (1997c). Chapter 5, Sections 5.1 to 5.2 and Chapter 1, Section 1.3 contain ideas discussed earlier in my PhD thesis and with more precision in Wissenburg (forthcoming). Chapter 6, Sections 6.1 to 6.5 are based on a paper I presented in Keele, and I used ideas from a paper presented in Florence to compose Chapter 7, Section 7.6 and Chapter 8, Section 8.1.

ACKNOWLEDGEMENTS

I estimate that more than 200 people have read at least one of the papers out of which this book evolved, one of the articles into which these papers were permutated, or one or more of the chapters in an embryonic phase. I cannot possibly thank them all individually, and I hope they will not feel insulted if I thank them collectively: the participants in the Environmental Sustainability and Social Justice seminars at Keele University (February, June and November 1996), the workshop on European Constitutional Decision Making at the ECPR Joint Sessions of Workshops in Oslo (March/April 1996), two workshops on liberalism at the Sixth International ISSEI Conference in Utrecht (August 1996), as well as those in workshops at the IRNES annual meeting in Lancaster (August 1996), the Irnes European seminars in London (April 1996) and Florence (June 1996) and the Environmental Justice Conference in Melbourne (October 1997). In this connection, I should also thank all the members of the Department of Political Science, now School of Public Affairs, at the University of Nijmegen. Over the years, I have had ample opportunity to learn to appreciate the really unique atmosphere, *esprit de corps*, support, flexibility and cordiality offered or exhibited by them all.

Some are more equal than others. For various reasons and on various occasions, discussions with the following people have had a more than average impact on the ideas presented here: Brian Barry, Milja Bos, Bob Goodin, Masja Nas, Oscar Struijvé, Sasja Tempelman, Martin van Hees, Albert Weale and Astrid Wissenburg. I should also mention those friends who offered me a chance to escape my obsession, friends with whom I hardly ever discussed my book: Ellen Hulsebosch, Hester Moerbeek and Marcel Verhoef.

The influence of five people went so far as to make them necessary and absolutely irreplaceable conditions for the completion of this book: Grahame Lock, who supported me without hesitation both professionally and personally; Andy Dobson, without whom I would never even have conceived of writing this book, let alone written a word of what I wrote; Marius de Geus and John Barry, who did a tremendous job reading every single line of the manuscript and taking every opportunity to point out the error of my ways; and Caroline Wintersgill for her comments and support, for leading by example (her energy seems to equal that of any electricity plant), for believing in the project and for making me sign with UCL and then proving I made the correct choice.

Finally, I am grateful to André Mahieu and his colleagues for their hospitality, for an almost uninterrupted supply of port and vodka, and most of all for supplying coasters in times of need, i.e. when I ran out of paper. While trying to relax and keep a safe distance from my project, I solved many a conceptual and compository problem in Café Mahieu and discovered numerous others there.

The usual reservation applies: none of the views presented here can or should be attributed to any of these 200 people, unless of course they claim the honour for themselves; the responsibility for representing and misrepresenting their views and those of others is mine and mine alone.

Marcel Wissenburg
Nijmegen, October 1997

Introduction

Imagine that we run out of oil; nothing will move. Or imagine that the ice caps start to melt, the land surface diminishes, the danger of floods increases, that people are being forced to move to higher places and others forced to live with them in a slowly dwindling territory. Or imagine the extinction of birds of prey. Mice, rats and rabbits, and with them pest, myxomatosis and rabies, will prosper like never before, until they have eaten everything on the menu. They leave a barren land behind where they – and we – starve. These are but some of the impending disasters green activists have announced if we do not change our evil ways. Nature being a whole, no part of it or plant in it being an island, everything being interconnected and interacting, every input immediately translating into output and side-effects all over the place, some have even predicted the occurrence of all these disasters and more within a few generations, ultimately resulting in Mother Nature striking back and exterminating humankind itself.

We do not have to believe in an ecological Armageddon to consider it possible that we may be on the verge of an ecological crisis. We do not have to believe in an ecological crisis to believe in the existence of a long series of ecological problems. And we do not even have to believe in these problems being *ecological* to believe that they are problems. If they exist in the minds of humans, they are at least moral problems, problems relating to the question of what we value and why we value it. One of the things we value is liberty or freedom: freedom of speech, freedom of thought, freedom of choice, freedom to have children, freedom to lead the life we want to live, and consumer freedom – perhaps one of the most underrated freedoms. These are freedoms warranted, protected and sometimes produced by the liberal democratic systems in which we live.

Ecological problems, it has often been contended, are problems of scarcity and hence of social justice. The scarcer a good is (believed to be), the less freedom there is to be distributed, or the lower the number of people who can profit from it. Ecological problems put limits to freedom, materially as well as theoretically; they put limits on the supply side of liberal democratic systems, though not necessarily

1

on the demand side. As a result of the perspectives that new technologies have opened up to us, scarcity is less than ever before a problem of the relative costs and benefits of goods and services, and more than ever one of their perceived physical scarcity. On the one hand, the possible freedom of consumers has increased beyond imagination: it is technically possible to satisfy more wants and desires than ever before. On the other hand, the environmental crisis confronts us with physical 'limits to growth' and with the costs of pollution in terms of money, health and freedom. As a consequence, liberal democratic political institutions are and will be confronted with a need to reconcile a growing number of momentarily satisfiable desires with an in the long run decreasing number of opportunities for desire satisfaction. This process, which in due course and other things being equal is likely to result in deprivation and disappointment, poses a fundamental threat to the survival of liberal institutions, as these are predicated on the possibility of reconciling antithetical preferences. If political systems can no longer deliver, or if people believe they no longer can, the systems may collapse. This is one among many reasons why ecological problems are interesting for theorists of liberalism, even if those problems exist only in the minds of citizens.

The flip side of the coin is that both the perception of environmental problems and the creation and implementation of environmental policies may in fact reinforce the legitimacy of liberal democracy. The prospect that the demands of citizens as consumers will become increasingly difficult to satisfy may make life uncomfortable for politicians and civil servants, yet their demands as environmentally concerned citizens offer immense opportunities for politically guided conflict resolution and preference satisfaction. Then again, it has been argued that the proceduralism and pluralism inherent in liberal democracy may make it difficult to accept and implement typically ends-oriented green policies (Dobson 1996).

A less obvious but equally important reason why liberal democrats can be interested in environmental issues is that the perception of environmental problems can make us aware of the shaky foundations on which the legitimacy of property (whether collective or private) rests. One of the most striking examples in this context is the Lockean proviso, according to which a property right to a resource x may require that 'enough and as good' of x is left for others. Locke's proviso presumes that resources are by definition abundantly available, whereas modern variants of it presume that resources are intersubstitutable ('fungible', in the words of Norton 1996) or infinitely recyclable, allowing for abundant (in a sense infinite) compensation for stock reductions.[1] The limits to growth have brought these two (and other) defences of the legitimacy of acquisition into serious difficulties.

There are more reasons why ecological problems threaten the legitimacy and survival of liberal democratic institutions: think of the extremely demanding economic consequences that the recognition of obligations to future generations and animals will have, or of the even more demanding effects of obligations towards nature itself if its intrinsic value were to be recognized (as ecological Utopians demand).

On a totally different note, a further reason why liberal theorists should be more interested in the green agenda is because practice is, as usual, storming ahead of theory: liberal democratic societies are already initiating and implementing environmental policies. Unfortunately, we register a certain degree of underdevelopment in the existing dominant two types of policy-guiding theory (ecological Utopianism and ecological modernization; the two will be introduced in Chapter 2). Environmental policies often lack a long-term perspective on an environmentally sane liberal democracy (cf. De Geus 1996: 212). In so far as current environmental politics and policies are theoretically inspired, this inspiration nearly always results in a short-term piecemeal approach, and on the rare occasions that a policy is justified with an appeal to a more extended theory (cf. the introduction to the first Dutch *Milieubeleidsplan*, National Environmental Policy Plan, VROM 1989), the 'higher' theory can hardly be called political. One of the aims of this book, then, is to fill the gap between the short-term ecological modernization and the often politically unfeasible long-term ecological Utopianism approaches: it should result in a practicable long-term answer to the environmental challenge, and thereby allow a theoretical assessment of political actions for the shorter term.

Environmental care is a much larger and far less theoretical minefield for liberal democracy. Some pro-environment measures – taxes, obligations, prohibitions – already infringe on individual liberty and social equality or have actually improved them, and they will probably infringe and improve even more in the future. Others create new recipients of government care at the expense of older 'clients': slowly, the legal status of the animal grows above and beyond that of an edible product into, unlikely as it may seem, a kind of legal person.

For the moment, these few examples should suffice to give the reader some idea of the sense in which ecological problems are problems for liberal democracies, and of the reasons why political theorists and philosophers who defend liberal democracy (I shall refer to them from now on simply as 'liberal theorists') need to address both environmental issues themselves and the theoretical framework of those who 'frame' environmental issues: radical and less radical ecologists, environmentalists, preservationists, conservationists (to whom I shall from now on refer as 'green theorists').

A second but no less important aim of this book is to fill a gap in political theory. For the last 20 years, the debates on environmental problems and on the moral foundations of liberal democratic societies have developed independently, where a fruitful dialogue between the two is both necessary and possible; the invitations were handed out years ago (Singer 1988; Taylor 1993). With only a few exceptions, theorists of liberal democracy like Rawls, Nozick, Ackerman, Dworkin and Nagel have paid little or no attention to what is popularly called 'the environmental crisis'; green theorists (again, with exceptions) have tended to see liberalism, liberal institutions, theory and values, as dependent variables relative to the indisputable directives of The Crisis. Some (Ophüls, Hardin and Heilbronner) have even seen liberal democracy as a cause of the problem rather than as an inadequate solution strategy (Doherty and De Geus 1996: 1–2).

I want to approach this issue from the opposite side: I want to find out how 'the' environmental crisis should be adapted to liberal democracy. Posing the question like that is impossible without some prior assumptions. I assume, first, that liberal democracy and liberal democratic values, including a commitment to freedom of choice and even consumer freedom, are here to stay. I also assume that a clear distinction can be made between political liberalism (liberal democratic political institutions) and economic liberalism (the free market as part of civil society); the two do not necessarily go hand-in-hand (cf. Lauber 1978). I focus on the former and discuss the latter most of the time merely as a sphere that is to be protected by and against, rather than interfered with by, the political institutions of liberal democracy. Next, I assume that 'the environment' is a social construction just like 'femininity', 'the economy' or 'liberal democracy'. More precisely, I assume that 'the environment' and political concepts relating to environmental issues like sustainability are mixtures of on the one hand empirical observations about what is the case and on the other moral assumptions about what is valuable and what ought or ought not to happen. I further assume that *is* (say, 'the' crisis) does not imply *ought* (say, unliberal values and policies). Finally, I assume that any ecological policy must aim to be feasible.

In my attempt to prove the possibility of green liberalism, 'assumptions' also play an important methodological role. I intend to proceed step by step, from one problem to another, assuming at each moment that other problems are solved, non-existent or 'constant factors'. Thus, when I discuss, for example, the issue of population policy, I initially assume closed borders (no immigration or emigration), a constant and uniform diet for and size of all individuals, a constant stock of natural resources, the absence of changes in technology and so on. In due course, all the factors that are kept constant at one time or another will become variable like they are in real life – but we cannot discuss all problems at the same time. The method I use to present my 'possibility theorem' reflects, up to a point, the analytical and deductive approach of rational and social choice theory, game theory and welfare theory (a classic example: Sen 1970). To keep this book legible, I have tried to find a golden mean between this style and the informal style of normative political theory by being far less formal and rigorous, both in structure and language, than rational choice theorists tend to be. No doubt this will annoy some members of both parties.

The first three chapters of this book are mainly descriptive and will be relatively tough reading, especially if the reader expects me to turn to things green and flourishing immediately. I will not. There is a lot of abstract theoretical work to be done first. In Chapter 1, I discuss the essentials of liberal democracy and formulate criteria for the moral acceptability of (environmental) policies from this liberal point of view. Although my list of criteria is long and complex, I shall, in the remainder of the book, usually concentrate on just three very broad characteristics of liberal democracy: the numerical equality of citizens; the proportional equality of recipients; and the liberty to live our life according to our own plan of life and our conception of the good. A fourth criterion, democratic decision-making as the expression of the ideal citizens' will, plays a crucial supportive role. Up to, and with the exception of, the last chapter, I shall take the citizens' preferences as a given and assume that they are as 'considered' and 'ideal' (though not necessarily as 'green') as preferences can be, the reasons being on the one hand that there is no prima facie reason for liberal democratic political institutions to interfere in the formation and transformation of citizens' preferences, and on the other that we want to know if liberal democracies can deal with green issues regardless of whether citizens care. Liberal democracy does entail a positive duty to protect and serve civil society as the sphere where public debate and private reflection shape preferences, thus transforming formal political democracy into deliberative democracy – but that is an aspect that will be dealt with in Chapter 9.

The first part of Chapter 2 contains a vocabulary of green political and philosophical ideas. Since it is not my aim to adapt liberalism to any one currently fashionable view on ecology or sustainability but to develop a liberal view on environmental affairs, we need a detailed taxonomy of all the green positions we can imagine to be possible, rather than a survey of existing schools. In the second part of this chapter, I limit the choice set, i.e. I deselect from the possible green theories those that are either intrinsically logically inconsistent or that are by definition incompatible with the liberal democratic values sketched in Chapter 1. This will still leave us with an impressive list of possibilities to choose from.

The third chapter deals with the (pre)history and future of green liberalism. I give a very short sketch of the conceptions of nature and environment that were used in earlier stages of the evolution of liberal democratic thought, thus at least in part explaining why or how liberalism seldom encountered ecologism. On the basis of this discussion, I introduce a series of 'green' issues that will, I expect, be part of the green agenda of liberal theorists in the next decades. I then discuss a few practical environmental problems that are on today's political agenda and that may pose problems in the realm of principles for liberal theorists.

Chapter 4 discusses the metaphysical and metaethical aspects of the value that liberals can or should attach to nature, to the environment and to its constituent parts. In more simple words, I ask here why we should care about the environment at all. I shall dismiss all classical radical green perceptions of nature as having intrinsic value, ultimately not because the concept of intrinsic value is defective or devoid of meaning, nor because it is obscure, but because it does not add anything helpful to practical discussions of environmental issues. I also exclude the possibility that nature has value for entities larger than individuals – say, species or biosystems. Instead, I opt for an understanding of nature as having external or instrumental value to any individual creature that can be conscious of harm and benefit. Since this obviously includes animals, I also discuss the possibility of animal rights here.

Chapter 5 deals with rights as such and focuses specifically on property rights. I develop an understanding of formal rights that allows us to see nature both as in principle possessable and as protectable. Moreover, it allows us to distinguish private and collective, attributed and unattributed, absolute, inalienable and limited rights, and to top things off, it allows us to represent any view of the good, any view on value and any plan of life. The gist of the argument made here is that rights cannot exist without being recognized, at least in principle and at least by reasonable representations of the people we actually are; that rights can only be recognized as such on the basis of good arguments, making property rights fundamentally limited; and that this implies the subjugation of property rights to the Restraint Principle. This principle demands, roughly, that no goods shall be taken from nature for the sole purpose of destruction; that they should not be destroyed unless that is unavoidable; and that if destruction is unavoidable, the original good should be reshaped, replaced or substituted for. The effect of this is the same if not better, from the radical green point of view, as claiming intrinsic value for nature.

Chapters 6, 7 and 8 deal with issues of sustainability, this currently being the dominant instrument of interpretation of, and dominant approach to, environmental problems. I distinguish three main roads towards sustainability: control of the demand side or consumption, the kind and volume of use that is being made of nature; control of the supply side or production; and control of the distribution process. In Chapter 6, I investigate whether liberal democracy can allow environmental policies aimed at demand-side control. In so far as this would mean volume control, i.e. influencing the size or number of individuals, their diet, taste or appetite, there is a clear conflict between control and the liberty of life. Moreover, the remaining alternative of control over the size of the population, which as we shall see is in fact a *conditio sine qua non* of guaranteed sustainability, turns out to be incompatible with any conception of society as a just scheme of cooperation. Any policy in this field will necessarily punish those who do not contribute to possible overpopulation and reward those who do. I call this instance of inevitable unfairness the Rapid Reproducers' Paradox. It is only in the realm of preference formation that liberal democracy might allow a form of demand control in the sense that it may ask for and assess good reasons behind claims to resources – a topic which I pursue in Chapters 8 and 9.

On the subject of a just scheme for the distribution of resources and products (Chapter 7), I observe that this is both the least and the most promising approach to sustainability. On the one hand, a just (re)distribution of benefits and burdens, whether global or society-wide, does not in any way necessarily influence the only relevant parameters, supply and demand. A more equal distribution of cars among humankind, to give a naïve illustration, is not the same as reducing the number of cars; a redistribution of cars may reduce stress on the environment in some places and it can but does not *have* to result in a globally tolerable level of pollution. Yet attention for the principles of distributive justice can be helpful in one respect: the development of new standards for good reasons behind claims to resources can possibly limit the problem of sustainability more than any other approach. In naïve terms, it may reduce the number of legitimately used cars. Yet it cannot offer any final solution to environmental problems.

As to supply-side control (Chapter 8), I only discuss those options that the sustainability discourse's model of the world as a nearly closed system allows: further exploitation of the Earth's resources, more efficiency, recycling. Although practice may be different, I argue that theoretically liberal democratic values are totally at odds with unrestrained supply-side solutions to the question of sustainability. In accordance with the restraint principle, liberal democracies are instead bound to allow exploitation of the Earth only to the degree that justifiable and recognized claims can be rewarded. Hence, policies directed at regulating and even diminishing production are admissible. This may not protect society from scarcity but it does protect nature from overexploitation.

Chapters 6 to 8 add up to the conclusion that liberal democracy is compatible with but cannot guarantee sustainability – no more nor less than any other political system. Yet a sustainable liberal democratic society need not be a beautiful place to live in: in a worst-case scenario, it could evolve into a global Manhattan without the Park and still be sustainable. If we want liberal democratic societies to be greener, we must look beyond the borders of the purely political and beyond the popular but unexplicit notion of sustainability.

In Chapter 9, I discuss several 'boundary-crossing' issues. First, I now make a variable out of what I assumed to be a constant factor throughout the book: the formation of individual preferences. Liberal democracy is totally incompatible with attempts to dictate peoples' tastes and preferences, yet we may reasonably assume that preferences are one of the determining factors of sustainability. On the other hand, and although liberal democratic institutions have no business interfering with civil society as the sphere where preferences are formed and transformed, it does have the task of protecting and preserving this sphere. I shall therefore investigate how far the combination of liberal democratic political institutions with a flourishing civil society (making it into what is called a deliberative democracy) can further sustainability. Second, I discuss the relation between political liberalism and economic liberalism, as one possible explanation for the by now manifest gap between the theoretically possible and morally prescribed sustainable liberal democratic society on the one hand, and the green perception of liberal democratic societies as destroyers of worlds. I point to some of the economic and cultural factors that will make a liberal democratic version of the sustainable society difficult, though not impossible, to realize.

1

Liberal democracy

1.1 Introduction

Since it is my intention to design a 'green' liberalism, a liberal democratic view on nature, the environment and environmental problems, the first thing to ask is: where to start? What precisely is liberal democracy about? The aim of this chapter is to develop a relatively formal model of criteria of liberal democracy. The criteria I formulate define a stringent but minimal version of liberal democracy; stringent in that the conditions are quite strict, but minimal in the sense that adapting them would either change nothing about our conclusions or would disfigure liberal democracy beyond recognition.

The first thing to note about liberal democracy is that it can be represented, like (to some degree) any other political system, as a system designed to translate preferences into rights. More precisely: it can be represented as the existence, on a collective level, of mediating and reconciliatory mechanisms (state, market, education, etc.), transforming the *claims* of individuals to benefits and the reduction of burdens into formal *rights* via the recognition of valid claims by means of *principles* of social justice. Claims can be seen as based on *preferences* and metadesires, and these in turn are intrinsically related to the individuals' *plans of life* and views of the good. Despite the diversity of conceptions of the good life we may assume the existence of a rough consensus on the need for these mechanisms – that is, a rough consensus on the institutions defining liberal democracy.

This is, obviously, a very one-dimensional view of liberal democracy and of politics in general. In real life, the decisions to translate some preferences of some citizens for, say, the conservation of a forest into a specific type of right for all, i.e. the obligation not to destroy the forest, in fact shapes or rearranges preferences. In this case, the lives, possibilities and perspectives of building constructors, businessmen and women, investors, nature lovers etc. are changed, hence their respective sets of (feasible) preferences are altered. Ideally, even the public debate on whether or not to protect a forest can change preferences: opponents may

9

actually convince one another of their views or influence them just enough to build a compromise. In short, it is not realistic to take preferences as a given. Yet this is precisely what I shall do, for the most part.

It is important to make a distinction here between the sphere of the political and the (super)sphere of civil society. Imagine an individual as taking part in dozens of types of interaction: as a citizen, she makes up her mind about political issues (or not) and votes (or not); as a member of a family, she is a parent, a child, single or not; as a consumer, she buys this or that here or there; as a producer, she works in a specific place and function or not. Each of these types of interaction between individuals has its rules and functions; they constitute 'spheres' of interaction. Interaction in one sphere can influence interaction in and even the rules of other spheres. In many societies, it is the business of the political sphere to authoritatively determine the rules and borders of all other spheres, loosely described as civil society. It 'overrules' the authority or capacity of any second sphere to determine the rules of a third. In liberal democratic societies, the task of the political sphere is further specified as that of protecting, as much as possible, (the spheres of) civil society against interference even from the political sphere itself. Exceptions are possible, thanks to the 'as much as possible' proviso.

The interactions within civil society, the views and preferences which are formulated there and which function as the input of the political sphere, can be seen as a second dimension of liberal democracy. The combination of both, the political input–output machine and the process of preference formation in civil society, creates a chain of reactions, a never ending dialogue between politics and civil society: the deliberative society. (There is a close link between this concept and that of 'deliberative democracy' (Doherty and De Geus 1996): if the deliberative society meets certain criteria of formal and actual intensity, it becomes a deliberative democracy. See Chapter 9.) Nevertheless, an *analytical* distinction can be made between the formation of preferences in civil society and the input of preferences in the political sphere for each specific instance of political decision-making, for each 'issue'. Now whether liberal democracy is compatible with green concerns depends first and foremost on whether the rules of this political sphere itself allow green politics to be considered, designed and implemented. If the rules do not allow this, green preferences are irrelevant because they simply cannot be translated into the appropriate rights. In that case, the rules of the political system may well be adapted to create the desired room for green concerns – but then it is no longer liberal democracy, at least not as we knew it before. To determine whether liberal democracy and environmentalism are compatible, we must therefore assume that preferences are given. Only after this issue has been settled, for better or for worse, can we talk about the desirability of changes in the liberal democratic political system and about the degree to which this would transform liberal democracy into something else.

In the following first two sections, I shall describe the input and output of liberal democratic systems, i.e. preferences (Section 1.2) and rights (Section 1.3). I shall give (more) special attention to the question why individual preferences would be so important as to make them the basis of political systems, to the genesis of preferences and to their incorporation in plans of life. Section 1.3 introduces a formal conception of rights that allows us to represent all sorts of plans of life, views of the good and measures of just distribution. In Sections 1.4, 1.5 and 1.6, the criteria describing the liberal–democratic understanding of respectively liberty, equality and democracy will be introduced. Some of these sections, the ones marked with an asterisk (*), are of a rather formal and technical nature. For the reader who does not feel qualified to read them or who could not care less, I have provided a hopefully more legible summary at the beginning of these sections, so that they can skip the rest without serious discomfort.

With the three sets of criteria from Sections 1.4 to 1.6 in hand, we should be able to assess ethical and political theories of the environment and environmental policies themselves, provided that we have reason to believe that liberal eyes are at all fit to see environmental problems. Whether this is the case will be discussed in the seventh and final section.

1.2 Preferences

As explained in the previous section, liberal democracy can be represented as a particular type of input–output machine, transforming the preferences of individuals into rights for individuals. Now 'preference' is quite an ambiguous concept: a preference can be informed or uninformed, it can be interpreted as a desire for a certain mental state or for an actual state of affairs and so forth (cf. Kymlicka 1990: 12 ff. on the notion of utility). Each interpretation presupposes a different view of the individual and of what matters about the individual. In other words, the exact nature of the input of liberal democracy depends on the reason(s) why liberal democrats believe the individual human being to be so all-important.

The fact that the individual matters more than anything else in liberal democratic thought is obvious. Even if individuals are sacrificed in a greater cause, liberals believe in that cause because of what it ultimately means to other individuals. It is individuals who are either the vehicles of whatever is good or the good in itself, and it is precisely because they believe something like this that they find a principle like *fiat justitia pereat mundus* (roughly: let justice prevail even if the world must perish in the cause) perverse. This feeling is not an exclusive prerogative of liberal democrats. There is a more general justification for it, which starts with the intuitive notion that we, individuals, are conscious of our individual existence and attach a certain value to our existence. We find ourselves (individually) important. In other beings we recognize the same distinctive trait of individuality; if we care about ourselves for this reason, then rationality bids us to do the same for them. Yet all that this primitive notion of self-importance establishes is that individuals matter universally. If other things matter less, why does the individual count for so much?

11

The history of philosophy has given us many reasons to place the individual at the centre of moral thinking. In Aristotelian thought, for example, human beings have a specific *telos*, an aim to strive for, a nature to realize, an inborn manifest destiny. It is the human's task or even calling to realize its potential, hence to cultivate its inborn capacities to the best of its abilities. From a moral point of view, this leaves humans with very little freedom to act. The measure for right and wrong is one that is in a way external to them: their nature. The woman who lacks any talent as an artist but still tries to become a painter because that is what she wants to be, rather than the brilliant rocket scientist inside her, fails the test of life. If liberal democrats were Aristotelians, the kind of preferences they would take into account would be 'natural', that is, not the (actual) preferences of the painter but the (potential) preferences of the rocket scientist.

Liberal thought assumes that humans are born free – at the very least meta-physically, and in one of three senses. It assumes that we can consciously deviate from the path of life that our inborn nature prescribes, if there is such a thing; it assumes that if we cannot deviate from that path, there is no sense in distinguish-ing between the actual and the potential because the two are then identical; and it assumes that if there is no such thing as an inborn nature, the whole category of the potential disappears. Hence, individuals are the makers of their own fate, or – since they may be metaphysically but not physically free – at least of their own plans.

Liberal thought assumes that man is born free and equal: all humans are equally free, no one makes another's plans or dreams or hopes. Again, there are ordinary physical limits to this. In everyday life, we communicate and interact and so change and shape one another's life. Yet basically, we all remain equally free to create, and equally responsible for, our plans.

In liberal democratic thought, then, the individual counts for more than any-thing else because she is the source of acknowledgment of all values, including the reason why one single individual x feels she matters. There is no external stan-dard to judge that fundamental reason by, except the reasons other individuals have for valuing their own individual existences – but these cannot overrule x's own standard since x is assumed to be free.

It is a commonplace to say that the things we find important about ourselves vary from individual to individual. We care about who we are, what we are, where we are, we care about our values and ideas, our thoughts, our products, the whereto of our lives. In liberal political theories, all this is caught under one heading: the individual's plan of life. It describes what an individual defines as her idea of a good life (technically known as 'the theory of the good'), how she wants to achieve this, and how she wants the rest of the world to accommodate to her plans. Plans of life vary not only from individual to individual but also in stability. They develop and change in the course of a life, becoming more or less internally consistent, more or less realistic, or simply different. They also vary in completeness. They can be limited to basic ideas about how we want our lives to go or our situation to change, or they can be extended further to include ideas about how we want our children to be and behave, our neighbours, our community, the world, the future of the world.

A plan of life can be represented or expressed in many ways: as a series of values and norms, as desires, as elections or in the form of preferences. The last may involve 'summarizing' a plan of life into comparisons between two or more alternatives; ultimately, a plan of life can also be represented, albeit in an inhumanly extensive form, as an ordering of preferences over all possible worlds. The choice for any specific way of representation is purely a matter of circumstances. In the context of a moral debate, norms and values are probably the most appropriate currency to work with; in an attempt to sympathize with a person, a novel or autobiography may also do; and on the psychoanalyst's sofa desires can be a good starting point. In liberal democratic politics, as I shall argue, we *can* talk in terms of preferences.

Let me first elucidate the concept of preferences. To translate human plans of life and the theories of the good from which they are derived into preferences, we must offer choices between *alternatives*, in a more technical than everyday sense of the word. Alternatives can be (sets of) two or more simple things x_1, x_2, ... x_n, like acts (cutting down a tree or not), goods (two distinct trees), even states of mind (anger or joy, etc.). They can also be sets of sets, as when we compare one nation's forests with those of another. At the most complex level, x_1, x_2, ... x_n stand for complete social states: that is, complete descriptions of the world including its history. In the last case, we could, for instance, compare two identical worlds that differ only with respect to the origins of Yellowstone Park: in one version of the history of the world it would have been planted by humans, in another grown naturally. The one thing that distinguishes technical alternatives from everyday alternatives is that the former must be of the same order. An act is no alternative to a social state; the two are incomparable. We do not choose between cutting down a tree and a world where that tree still grows, we choose either between two worlds with and without the tree, or between cutting the tree down and not cutting it down.

Preferences are expressed in the form of *orderings* over (sets of) alternatives: x is at least as good as y – the formal notation is xRy. If x is at least as good as y but y is not at least as good as x, we strictly prefer x over y: xPy. Finally, if x is at least as good as y and y at least as good as x, we are indifferent: xIy.

It has been argued, among others by Meir Dan-Cohen, that preferences are not just an artificial currency but even unrealistic. Humans would not look at the world in terms of preferences only. They do enter a restaurant, look at the menu, list dishes according to their preference from their most favourite food to the garbage they dislike most, and *select* their number one choice – but they may sometimes simply look over the menu and *elect* something. The latter situation is comparable to that of a sculptor who, after the act of creation, explains the resulting sculpture as the form that was already 'in' the block of marble in the beginning, trying to get out. It also seems appropriate to discuss love in terms of election rather than selection: as a rule, we do not list all possible lovers and then choose to fall for Number One. Dan-Cohen argues that we choose on the basis of values and that values do not always admit of degrees of satisfaction: an item either meets the standard or it does not (Dan-Cohen 1992: 228). Hence, preferences cannot represent all the choices we make in life.

Dan-Cohen's objections are forceful, even more forceful than we might expect. They do not merely rest on psychological observations, although that might be convincing enough, but on philosophical arguments as well. Election, he argues, is at least as compatible with conceptions of moral autonomy as selection, and sometimes it offers an even better understanding of choices. For example, election is a good representation of the Kantian ideal of moral autonomy: it expresses the 'inner necessity' of a will guided by a moral maxim in ways preference orderings do not (Dan-Cohen 1992: 233). No Kantian would rank slaves according to the degree to which they deserve freedom since the categorical imperative simply excludes the admissibility of any kind of slavery. Arguments of this sort are especially appealing to liberal democrats who, as we saw, value moral autonomy and individual moral responsibility so highly. Nonetheless, accepting that election may be a better description for some of the choices we make in life does not make descriptions of those same choices in terms of preferences impossible. It shows that a preference for the continued existence of a rainforest over its being transformed into mahogany toilet seats may be a matter of absolute rejection of one alternative instead of one of stronger and weaker inclinations. Yet we can still talk of a preference for the one over the other. A translation of people's values in terms of preferences does not mean that we cannot, if necessary, attach measures for the intensity of a preference to alternatives (in this case, absolute zero to the toilet seats and one to the pristine forest) or, alternatively, give individuals a veto power when it comes to collective decision-making. Preferences offer a way to construct a model of the individual's likes and dislikes. The model may be artificial but there is no reason to assume that it must also be unrealistic.

Now let us assume that individuals know exactly what is good for them, now and in the future, that they order their preferences accordingly and voice them honestly. I shall drop this assumption in a few moments, but I need it now to show how politics and preferences are related, even in a world of such nearly perfect individuals. We take ourselves, our existence and plans seriously; hence, we do not immediately subject ourselves to another's will – to the first one who speaks, the strongest, the early riser. At this point, politics is born: either we cooperate or separate. In the process of designing and redesigning a viable scheme of cooperation, we have to take account of all individuals, at least in principle, as well as of the things they care about. Thus, we must acknowledge the existence, on an individual level, of preferences and, particularly, of a metapreference for freedom of choice, i.e. a desire to be able and permitted to satisfy our own preferences. The precise content of this metapreference:

(a) depends on an individual's conception of the good or desirable life as expressed in a plan of life

(b) can consist of all imaginable kinds of preferences: ordinary first-order preferences (what I like for myself), second-order preferences (what I want others to do), etc.; and

(c) it can take any form.

The last condition allows anything from a desire for freedom to form our own preferences independently, down to one for a very limited freedom of preference-building guided by authorities or an authoritative substantive ethics, or even to the total denial of freedom and the submission to another's will – what in some circumstances might be called altruism.

Like all political systems, liberal democracy focuses on reconciling preferences in some way: on ordering individual preferences, determining which combinations are feasible and turning the most desirable possible world into reality. Obviously, this requires standards for ordering, feasibility and desirability – but prior to that it requires a standard for admissibility, for the recognition of preferences as valid claims. Even in liberal democratic societies, not all preferences are welcome or can or will be considered. My possible preference for world dictatorship certainly will not.

It is at this point that other important objections against the concept of preferences and its applicability have been raised. We can discard four of these as relatively insignificant: individual moral pluralism, intrapersonal measurement, intersubjective measurement and incompatibility. The first of these objections states that, in assessing alternatives, individuals use diverging criteria, values that cannot be reduced to one another nor lexically ordered. They judge, for instance, art by beauty and/or by compatibility with the colours of the curtains, or the physician's performance by their health and/or by the impact of her comforting words. Both on an individual and on the social scale it would be difficult to transform all these judgements into one complete preference ordering. The second objection is closely related to the first: even if they have only one criterion, individuals often cannot or will not make up their minds about their preferences, let alone the weight they attach to alternatives. Third, there is no guarantee that the intensity of my preference for alternative x matches your dislike for it, even though x is at the top of my list and at the bottom of yours. Fourth, some people's preferences are simply incompatible.

The problem with these four very real objections is that they will not go away if we drop the idea of preferences. They apply, *mutatis mutandis*, in the same degree to all other ways and means in which political systems try to take account of the will of the people. They are problems for standards, not for the objects (preferences) to which the standards are applied. The way in which liberal democracy tries to deal with them, the criteria it sets for liberal democratic decisions and decision-making, will be discussed in subsequent sections, where I implicitly answer these objections.

A more relevant objection to my use of the notion of preferences so far would be that some preferences do not deserve the name of preference, exactly because some people do not know what is good for them, do not order their preferences according to their own standards or voice them dishonestly. One possible rejoinder would be to argue that whether or not uninformed, erratic or false preferences should be disregarded depends on the rules of the political system for the recognition of a given preference as a valid claim – but what is in a name? The problem of whether preferences should be informed, consistent and honest remains.

Liberal democratic political systems have a prima facie duty not to interfere with the genesis of preferences. People's plans of life, including their theories of the good, matter and matter equally; hence, the theory of the good and whatever other criteria an individual uses in ordering her preferences matter. Their freedom to formulate plans and preferences is a precondition of morality and implicitly of the moral status of a society: no individual can be said to act (im)morally if she acts under another's control. It follows that a liberal democratic system has a prima facie duty to consider all preferences, the right as well as the wrong ones.

Duties may clash. We never refer to a duty as prima facie except to announce exceptions. Preferences can be uninformed because of deliberate misinformation by others, they can be inconsistent because of a lack of experience or education, they can reflect class-based prejudices, they can be dishonest because of bribes, they can be simply physically infeasible (compare my possible world dictatorship with my impossible dictatorship of the Roman Empire). These and others are certainly good reasons to interfere in the genesis of preferences, exactly because non-interference would contradict the spirit of the prima facie duty.

However, once a political system itself, rather than individuals or collectives in civil society, starts to question the individuals' preferences because of the theory of the good on which they are based, we enter a twilight zone where political institutions risk becoming puppet masters. It can be argued that even theories of the good are not beyond impartial criticism since they may be, for instance, internally inconsistent or unreasonable (cf. Brandt 1979; Rawls 1993; Wissenburg forthcoming), but this is probably as far as state interference in preference formation can go without in some way or other inhibiting rather than protecting moral autonomy.

A final objection against using preferences to represent individuals' plans of life is derived from Bryan Norton's critique of popular interpretations of preferences in economics (Norton 1994). It relates to the idea that an individual's preferences should not simply be taken as given because they *are* artefacts, created, at least in part, by others than the individual. Preferences arise out of reflection and debate with or the influence of others in society. The objection is not that preferences (or theories of the good, for that matter) cannot be taken to represent the 'real' individual expressing them, but rather that they only represent her at a given point in time. Preferences cannot be assumed to be fixed and unquestionable (cf. Norton 1994: 320) – the fact of the matter is that they change because they are questioned. Whatever preferences may represent, they cannot represent 'the' good of the individual. Hence, preferences cannot be the whole truth about what matters about and to an individual, nor can they serve at the sole point of departure for an analysis of a political system.

Part of this objection is correct, part of it is not. It is true that there is no such thing as 'consumer sovereignty', as Norton calls it. It is true that both preferences and the theories of the good behind them are unstable, it is true that they change in the course of a life or a day because of deliberation and persuasion, it is probably true that they cannot even develop or exist without a social context enabling the individual to shape them. But it does not follow that they cannot represent an individual, nor that they cannot be taken as given in our context. For starters, *taking* preferences as given does not imply that they *are*. Here, it only implies that preferences are not something a liberal democratic political system will frolic around with. As explained in the beginning of this chapter, the genesis of preferences is a matter for civil society. In liberal democracies, one of the tasks of the political system is to establish and protect the rules of interaction within and between the spheres of civil society in a way that promotes moral autonomy. Questioning and (authoritatively) mutating preferences *at* the moment of decision-making would in fact annihilate the role of civil society. Obviously, if we admit that preferences are mutable and at best represent the individual's good at a single point in time, they cannot represent the individual's good as such. There is no way in which this can ever be done. An individual's past preferences are no longer relevant, her future preferences are unknown. The only way in which an individual's good can be represented by preferences is if we take her current preferences as the input of a political system at that *one* point in time.

To end with an illustration: imagine a society in which, among others, liberals and greens live. At a given point in time, an individual *l* has a preference for, say, possessing a mahogany toilet-seat over prohibiting the import of mahogany, whereas individual *g* has a preference for protecting the lungs of the earth over cutting down trees to satisfy extravagant desires. Deliberation within the boundaries of civil society may change the preferences of either one of them and may even lead to both sharing the same preference for *l* not buying a mahogany toilet-seat. However, *l* discovers that she also has a preference that outweighs every other mahogany-related preference: she wants to be free to decide whether or not to buy mahogany products. She wants it to be her own decision, rather than one taken for her by others. By a quirk of fate, it is at this precise moment that 'the political system' is supposed to decide on a ban on the import of mahogany. 'The system' lists current preferences and discovers that *l* and *g* have incompatible preferences. It takes a decision and life goes on as before, as does the dialogue between *l* and *g*. In due course, their preferences change again and perhaps a new political decision is called for, at which time the process starts all over again. The question I want to address in this book is not how the interaction within civil society should proceed in order to get *l* and *g* to agree on preferences or to get *l* to accept *g*'s preferences, nor how civil society and the political system should interact in order to reach either of these results or to at least ensure that *g*'s preferences are enacted. What I am interested in are the rules used by the political system at the moment of decision-making. The question I want to answer is this: can the rules of a liberal democratic political system *themselves* in any way accommodate g's type of preferences for a greener world, regardless of the current preferences of the individuals who accidentally happen to be around? Because if they cannot, then anyone who believes that both a sustainable world and the values incarnated in liberal democracy are necessary conditions for a valuable human life has a serious problem.

1.3 Rights ★★★

This section is one of the two 'technical' bits of this book. Its aim is to show how the output of a political system can be represented, with a little help from deontic logic, in terms of so-called formal rights. For the benefit of readers who take the legitimate and respectable position that they do not have the time and energy for a crash-course in deontic logic, I open this section with a summary in plain English of the technical bits. They can then skip the part that starts under the mark ★★★ and move on to the next section, hopefully without ever feeling they missed anything.

As we saw in the previous section, liberal democratic political systems need standards to be able to order 'admissible' (legitimate) preferences: in other words, claims. In the course of a process of consideration and reconciliation, principles of justice, liberty, equality and so forth are used to order claims and perhaps exclude some as too illiberal, undemocratic or unjust. Subsequently a decision is made by means of a final principle, by a qualified or unqualified majority, unanimously or in some other way. What these principles are is the subject of the ensuing sections. Here, I want to concentrate on the output of the political process, which is a description of the, at that moment, most desirable feasible world. It defines the things people will and will not have or do, sometimes abstaining from a clear choice. This end product can be described in terms of duties, permissions, liberties, etc., all of which are instances of 'formal rights'.

What is a 'formal' right? 'Formal' means nothing more than 'represented in the form of formal logic'. Rights, in the everyday legal sense of the word, are necessarily part of the output of any political system. There is no public transport without a right to travel on buses, no private transport without a right to travel, no social security without a right to claim, no elected government without a right to vote, no dictatorship without an exclusive right to govern. In fact, all of a political system's output can be phrased in terms of rights and the limitation of rights. It is reasonable to want a system of rights not to create inconsistencies: if one rule demands a clean environment for each individual, and another prohibits the invasion of other people's property – say, a polluting factory – we need a rule to decide which of the two has priority. In designing such an internally consistent system of rules, logic is indispensable.

In the logical representation of rights I use throughout this book, rights are seen as permissions and negations of permissions. Initially, they apply to one individual only, at one place and one moment in time, for one 'basic act'. I shall explain that last term in a moment. For now, let us say that 'cutting down a tree' is a basic act. Now legal rules can be quite complicated. If I am permitted, at this very moment, to cut down the tree outside my window but not permitted *not* to cut it down, I have an *obligation* to cut it down. If I am permitted to cut it down now and permitted to *not* cut it down, I have a *liberty* to cut it down. If I am not permitted to cut it down but permitted to not cut it down, I have a *duty* not to cut it down. Things can become even more complicated when all the law specifies is that no one is to interfere with my cutting down the tree or not. The only substantial information this rule contains, as far as my duties and liberties are concerned, is that I do *not not* have the permission to cut it down and that I do *not not* have the permission not to cut it down.

Rules of this type – one person, one basic act, one point in time and space – will be called basic or *molecular* rights. They are dissimilar to everyday legal rights in that they can represent all sorts of (non-)permissions, rather than the one thing a legal right usually describes: the combination of a permission to x and a permission to not-x. The notion of a basic act plays a crucial role in basic or molecular rights. It is the unit of the measurement of rights. When we try to judge, say, a non-accidental death, we need to know which means were used, the switch of the lung machine or the trigger of a gun; we need to know to what purpose, for profit or for humanitarian reasons; and we need to know how it was done, with one's own hands or using those of others. In other words, when we judge the legitimacy of an act, we must be able to individuate it as one *single* act. Hence, we need to distinguish the smallest possible building brick of rights: one purpose, one means and one single act, an act that can be performed without any one specific prior act necessarily taking place. This is what a basic act describes. Reading a particular book to gain information is one basic act, reading it to pretend being engaged is a second, and using a particular computer and software for information purposes a third.

When rules cover more than one basic act and describe situations like the precise times or days at which I may or may not cut down trees, the exact trees or number of trees I may cut down, or who exactly is (dis)allowed to cut down this particular tree, things become more complicated. I shall therefore refer to the collections of molecular rights that describe these complex situations as *complex* rights. Legal rules are nearly always complex rights of this sort: they apply to whole classes of individuals ('citizens'), to complete regions ('the nation'), to a whole *genus* of basic acts (wielding axes, using drugs, having sex).

Representing rules in the form of legal rules, i.e. everyday-like, appears to have one advantage over representing them logically, i.e. as formal rights: legal rules seem more compact. We need (in principle) only three rules to solve the question whether or not I am allowed to stop a chemical plant from polluting the river I live on: one about pollution, one about invading private property and a priority rule. In terms of formal molecular rights, the response to that same problem would be an infinitely precise list of all my (non-)permissions to molest this particular object at any moment in time. The advantage disappears when a legal issue get more complicated; ultimately, we may well be forced to spell out permissions for nearly every single act. The advantages of thinking *in terms of* formal rights are far greater. First, formal rights are economical: they can represent all legal rules from prohibitions to policy goals with just two words. Second, formal rights are individual rights, whereas legal rules can be indeterminate about individual freedoms or obligations. Third, the language (and logic) of formal rights can be applied not only to legal rules but also to moral rules: that is, to our subject. Critics of contemporary political philosophy have remarked that the discipline suffers from a lack of precision; for example, when authors compare moral principles in terms of the freedom they offer but refrain from specifying exact measures of freedom (Carter 1995a: 819). Although I will not attempt to represent every one of the claims, rules and principles discussed in this book in the grammar of formal logic, I do hope that the use I shall make of formal rights and other formal instruments convinces the critics of the *possibility* of more precision in political philosophy.

★★★

My formal conception of rights is partly a simplification of models used in deontic logic and partly a sophistication of them (Kanger and Kanger 1966; Lindahl 1977; Van Hees 1995a; 1995b). On the one hand, I distinguish only a limited number of so-called types of right; many of those usually mentioned in deontic logic are not essential to our quest. On the other, I demand far more precision than usual as regards the description of the subject of rights, that is, the precise 'thing' or 'act' to which a right can exist. In informal terms, I understand rights as sets of various types of permissions. Permissions describe the legal or moral possibilities of an individual i to perform a basic act: that is, any single act that can be performed without any prior acts necessarily taking place. This definition is based on Steiner's (Steiner 1983); it is not uncontroversial but a discussion of the proper way to define a basic act would lead us too far astray. A basic act is described as doing exactly one thing x with exactly one thing y to exactly one purpose or reason z, at or during one given time(slice) t and at a certain place or in a certain area p (cf. Wissenburg forthcoming; Parent 1974). There are several types of permission: to do a or $\neg a$ (liberty), to do a (duty), to do $\neg a$ (duty), to do neither a nor $\neg a$, or more subtle possibilities like a permission not to do a or the absence of a permission to do a or $\neg a$. In what follows, I assume that *tertium non datur:* either we can do x or we can do $\neg x$, but we cannot abstain from acting. I also assume, not for reasons of principle but merely to keep things simple, that there is no practical use in talking about rights that we cannot physically perform or avoid performing. At best, permissions to jump over the moon or to use our hair to see are virtual rights.

Finally, I distinguish between molecular and complex rights. A molecular right assigns to one individual one permission for one basic act. A complex right is simply any set, any collection, of permissions with at least one element, one permission, in it. Note that (molecular) rights are assigned to individuals without making reference to either a moral standard or to the value or weight that any particular right would have according to that standard.

Formally, a right in the simple non-metaphorical sense is the undivided authority to decide (this phrase will be explained shortly), or the full control over, whether or not a specific X will be used in a specific way Y to a specific purpose Z. Against this background, we define:

X as a set $\{x_1, x_2, \ldots x_n\}$ of objects
Y as the set $\{y_1, y_2, \ldots y_m\}$ of means; and
Z as the set $\{z_1, z_2, \ldots z_l\}$ of ends.

Note that I do *not* assume that X, Y and Z are disjoint. A book can function as a means or as an object.

The basic material out of which rights are made will be called right-molecules or *r-molecules*. They define a person's physical and moral freedom, at one particular moment and place, to autonomously determine whether one particular object shall be used in one particular way to one particular purpose. In formal terms, an r-molecule is:

(1) an element (x_a, y_a, z_a) from the set $X \times Y \times Z$
(2) at one particular moment in time and on one particular place (t, p).

I make three assumptions regarding r-molecules:

(1) an r-molecule is something over which one and only one person i has sole and exclusive control, i.e. *i sees to it that* (cf. Lindahl 1977: 28) one of the two possible worlds will exist: the one in which i sees to it that (x_a, y_a, z_a) is the case at (t,p), or the one in which i sees to it that (x_a, y_a, z_a) is not the case at (t,p)
(2) r-molecules describe basic acts, i.e. those acts that require no one particular necessary prior act
(3) condition $\mathfrak{C}$, the condition of absence of instrumental obstruction: it is not true that it is physically impossible to influence (x_a, y_a, z_a).

This last condition requires further explanation. The absence of instrumental obstruction is what distinguishes a genuine right from a formality. When the Prussian King Frederick the Great in 1751 granted the *Recht auf Herrschaft* (sovereignty) over the moon to a certain Aul Jürgens, he did not grant any rights but performed an empty formality. Condition $\mathfrak{C}$ may be controversial. Its implication appears to be that rights cannot exist when they cannot be exercised – my right to clear water, for instance, would supposedly not exist as long as there was no water around at all, or if there were clean water but I was taken hostage and prevented from getting hold of it. However, these two cases must be distinguished. In the latter, someone *violates* my right, which is not the same as denying its legitimacy or creating a law of nature that makes it utterly impossible for me to

21

drink clear water. In the first case, in the absence of clear water, I still have a right to it: a right to clear water *as soon as* it is available. (Finally, for reasons of simplicity I also assume that there are no conditions other than $\mathfrak{C}$ inhibiting the exercise of rights.)

In summary, an *r*-molecule is a quintuple denoted as *i*: $\mathfrak{C}(x_a, y_a, z_a, t, p)$, or, shorter, as *a*, *b*, *c*, *et cetera*. An example of an *r*-molecule would be *i*: $\mathfrak{C}(x_b, y_b, z_b, t, p)$, in which x_b stands for this book, y_b for burning it page by page and z_b for the purpose of expressing outrage. *You*: $\mathfrak{C}(x_b, y_c, z_b, t, p)$ means that you can throw it out of the nearest window as well and *you*: $\mathfrak{C}(x_b, y_c, z_c, t, p)$ means that you can also do that to express confusion. All this at the appropriate time and place.

In everyday life, we accept that complex rules can be imprecise and incomplete: my right to use pesticides may interfere with or be limited by other rights. I am perfectly free to drink pesticides, bathe in them or offer them to a guest, as long as I do not use them to poison someone, which would interfere with other, higher rights. We assign generalized versions of *r*-molecules, and in everyday speech it is these complexes which we call 'rights'. They can be (nearly) all-purpose rights, for instance a right to read one particular book for (nearly) any purpose; or (nearly) all-object rights, such as a right to save a life at (nearly) any cost; or (nearly) all-means rights, such as a right to use a book in (nearly) any way to teach a class. A right can also be a complex of such complexes.

There are several types of rights. Remember that an *r*-molecule was defined as *i*: $\mathfrak{C}(x_a, y_a, z_a, t, p)$ and denoted simply as *a*, *b*, *c*. Rights are complexes of *r*-molecules. A *basic right* r_a is defined as a right over one single *r*-molecule *a*:

(1) r_a is the authority to decide that *a* and the authority to decide that $\neg a$; or
(2) r_a means that *it shall be the case* that *i* sees to it that *a*, or that *i* sees to it that $\neg a$.
(For further interpretations of these operators see Lindahl 1977: 52 ff.)

A basic *duty*, to describe just one other possible type of right, is a limited basic right, limited by taking away the element of choice. A basic duty d_a is the authority to decide that *a*, a basic duty $d_{\neg a}$ is the authority to decide that $\neg a$. Basic duties give us an authority to do something rather than to choose to do it.

A *complex right* is a set of one or more basic rights and/or basic duties, e.g. (r_a, r_b). Incidentally, the addition of more basic rights does not mean that our choice of alternatives grows to the same extent. Suppose that:

a is *i*: $\mathfrak{C}(x_1, y_1, z_1)$
b is *i*: $\mathfrak{C}(x_1, y_2, z_1)$
c is *i*: $\mathfrak{C}(x_3, y_4, z_5)$
d is *i*: $\mathfrak{C}(x_1, y_4, z_5)$.

A complex right (r_a, r_c) gives us four options, four possible worlds: (a,c), $(\neg a,c)$, $(a,\neg c)$ and $(\neg a,\neg c)$. The complex right (r_b, r_c) also gives four alternatives, but (r_a, r_b) only gives three: $(a,\neg b)$, $(\neg a,b)$ and $(\neg a,\neg b)$. In English, this means that we cannot have our cake and eat it.

Some complex rights, like the complex (r_a, r_d) in our example above, consist of basic rights with at least one common denominator, a shared x, y or z. A complex right is a *perfect right* $\Re x_a$, $\Re y_a$ or $\Re z_a$ if it contains all possible basic rights with regard to an x_a, y_a, or z_a. Suppose that (r_a, r_d) is the complex right to use a particular fruit knife either to stab in self-defence or to cut an apple with. If these were all the things we could do with that fruit knife, it would be a perfect right of the $\Re x_a$-type. Rights in the real world are always imperfect or *conditional* rights: in other words, perfect rights with a long list of exceptions.

By analogy, we can see a complex duty as a combination of two or more basic duties and a perfect duty as a universally valid complex duty. In ordinary language, the last would be expressed like this: 'whatever you use to whatever purpose, see to it that you use it only in this the right way.' The third version of Kant's categorical imperative, demanding that we should always act in such a way that mankind and each human are never merely means but also ends (Kant 1974: BA66), is a beautiful example of a perfect duty with two common denominators. Note, however, that complex duties are merely a special case of complex rights.

There is one kind of duty that our model cannot represent all that simply: the complementary duty others (can) have to respect or even protect the rights of individual i. As an r-molecule can only be assigned once, it cannot be the subject of one individual's right and of another individual's duty. A complementary duty would have to be expressed in terms of its implications: if I have, in practice, a duty to respect your freedom of speech, I have, in our model, a (complex) duty to see to it that certain things (r-molecules) that would physically obstruct your complex right to free speech do not happen.

This brings me to the last subject of this section: the fit between this model and conventions in deontic logic on the proper description of rights. We have already dealt with one difference in modes of expression: whereas my complex rights are defined in terms of an individual and certain objects, rights as understood by a leading deontic logician, Kanger, are relational: they describe the relation an individual has to another individual regarding objects (e.g. Kanger and Kanger 1966: 86). As I have just shown, the two models are compatible: the model I use *can* take account of relations by describing other people's complements of one individual's complex rights.

A more important difference between the conception of rights presented here and current conventions in deontic logic concerns the notion of *simple rights* as first described by Kanger (Kanger and Kanger 1966; cf. Lindahl 1977: 43). In Kanger's taxonomy, there are four types of simple rights from which other types are derived: a *claim* (X has versus Y a claim that $S(X,Y)$ means that: it shall be that Y causes $S(X,Y)$; $S(X,Y)$ describes a state of affairs between x and Y), a *freedom* (not: it shall be that X causes that not-$S(X,Y)$), a *power* (not: it shall be that not: X causes that $S(X,Y)$) and an *immunity* (it shall be that not: Y causes that not-$S(X,Y)$). It is a claim I cannot defend in detail here, but the conception of basic rights defended here and that of Kanger's simple rights are again compatible: instead of i: $\mathfrak{C}(x_a, y_a, z_a, t, p)$ we can read $S(i,j)$; the rest of the framework can be adapted accordingly.

1.4 Democracy

> British democracy recognizes that you need a system to protect the important
> things of life and keep them out of the hands of the barbarians. Things like the
> opera, Radio 3, the countryside, the law, the universities. *Both* of them. (Sir
> Humphrey Appleby in *Yes, Prime Minister.* 5: *Power to the people*)

Rights represent the stuff politics is made of. Institutions like councils, county
or provincial authorities, the state and its ministries, the European Union and its
Directorates, the United Nations and UN agencies are all involved in the great
political process of transforming preferences into rights. If they want to perform
this task, they must be viable, and in order to be viable political entities, they must
first of all be the supreme authority in the respective policy fields they cover. To
the extent that they make their own decisions, and thus limit the power of others,
they have an independent existence as *political* entities; to the extent that they
have no alternative but to carry out the decisions of (by definition) higher authorities, they do not exist as independent political entities. Were the (for example)
European government a committee of representatives of member states, each having
veto power (a not too Utopian scenario), Europe would not really exist – not as
an independent political entity. It would be less general and less relatively independent than the UN, it would even be less independent than a specialized
organization like the International Postal Union.

For the same purpose of viability, political institutions also require legitimacy
in an amoral sense of the word: support of those whose power matters. In a
democratic state, this means popular support; in a free market society, support
from producers, consumers and workers; in a *Rechtsstaat*, the backing of judges
and courts. There is at the background of all this room for legitimacy in the
moral sense as visible in an adequate protection of citizens against one another
and against government abuse of power, and in other traits usually associated
with legitimate governance. Yet moral legitimacy is only an indirect criterion for
viability, mediated by those whose judgement actually counts and for whom
presumably other criteria may be at least equally relevant.

Viable political institutions need a combination of two things: a constitution of
sorts to delineate their responsibilities relative to others' institutions (other states,
regions or citizens) and a social contract to secure its legitimacy. More precisely,
they need a constitution that is, at the same time, a social contract. Note that the
demand for viability does not imply that such a constitution must be unchangeable or even stable and unchanging, merely that it must be relatively stable.
Constitutional rules are not rules like all others – they are metarules: rules for
making rules, for making and protecting the rules that govern civil society. A
change in civil society on details, say one or two rules, need not endanger the
stability of the fabric of civil society, but if the metarules are unstable, *all* the rules
they determine become equally uncertain, equally provisional, momentary and
insignificant, leaving civil society with no structure worth mentioning. Thus,

even though constitutional decision-making goes on continually, there is good reason to create extra safeguards against changes in constitutions. Since constitutions do not usually drop from the skies on stone tablets, and since this might not always be the most prudent way to gain legitimacy, they have to be invented. This is where side-constraints other than viability can come into the picture. Assuming, for the sake of argument, that the notion of viability covers the whole range of substantive empirical conditions for constitutional decision-making (henceforth: CDM), we can distinguish two other categories: substantive normative conditions and purely formal conditions. The latter limit the procedures by which CDM can possibly take place to those that meet, say, common criteria for rational choice like transitivity, completeness, independence of irrelevant alternatives, neutrality (in the rational choice sense of the word) and so forth. But their importance is over-shadowed by that of the substantive normative conditions for three, fairly obvious, reasons. One is that we value procedures for more reasons than their rational or decisive qualities only – we can learn to live with logically imperfect procedures for the sake of democracy or representativity. Second, even the reasons why we value purely formal criteria are of a normative nature: transitivity, for instance, is not something we desire for its own sake but because we have certain ideas about the qualities that make our reasons for preferring x over y good reasons. Third, we often value procedures for the results they exclude or ensure. Procedures that exclude the possibility of invading property rights by collective decision mechanisms or dictators might win the favour of economic liberals more easily than those that do not. We can then, perhaps paradoxically, conclude that a necessary condition for agreement on *how* CDM is to take place, is a sufficient degree of agreement on *what* the outcome of the CDM process should be.

This brings us to the matter of the ultimately desirable content of a constitution. We can broadly distinguish between two major issues here. One is the existence and capacities of instruments designed to protect and uphold the political institution and its constitution – instruments like, in the traditional nation-state, courts of law, police, army, procedures to select rulers and so forth. The other is what I call the *substance* of the constitution, a set of three overlapping issues: the policy areas with which a political institution is concerned, the values which it is to protect or advance, and the priority order(s) in both areas. Against this background, it will be evident that one of the most relevant side-constraints or aims for many political institutions is to preserve the liberal democratic heritage. Specifically the nations of the West are products and guardians of a set of post-1789 values collectively labelled as 'liberal democratic'. These values are appreciated widely, even by the powers that matter. In the process of CDM, no new or old political institution can avoid decisions that will determine whether, where and how this heritage is protected.

We might argue that the values of liberal democracy are more than merely normative conditions, more than things certain people just happen to appreciate. On this account, they are (also) actual cornerstones of many societies, even defining characteristics. Had Western societies not incorporated traits like freedom of conscience and expression, freedom of movement and contract, social and legal equality, they would not have been what they are today or would not even have existed at all. Hence, the room Western institutions offer to liberal democratic values will be of more than average importance both for their viability and for the moral stature of their constitutions.

The trouble with liberal democracy is that it is a kind of hurrah-phrase – a vague multifunctional term with hearteningly positive connotations. Political scientists and theorists normally do not bother to analyse it beyond the point of remarking that it has something to do with democratic control over rulers, with civic liberty, and with equality in some form or other – legal, political, sometimes social. The theorist usually moves on to a formal or normative analysis of what either of these three concepts respectively should mean, the empirical scientist investigates their permutations in concrete cases. Even if it was not designed to fit the specifics of the European context and the case of environmental care, the characterization of liberal democracy that I am about to offer in this section and the next two would, therefore, still be destined to be idiosyncratic. Yet I do not believe that it is also implausible; it seems compatible with such classic statements of the liberal democratic ideology as De Toqueville's (1951), Paine's (1989), Publius's (1962), Dahl's (1956) and Rawls's (1972, 1993, 1995).

The three basic ingredients of liberal democracy (democracy, liberty and equality) are, to use another hurrah-phrase, essentially contested concepts: that is to say, no one doubts that these elements are essential to it but what they mean, exactly, is open to debate. To some degree, this controversy can be explained by the fact that we may be trying to combine irreconcilable conceptions: that is, interpretations of a concept. As we shall see, the concept of democracy is represented among other things by the conceptions of the will of both ordinary and ideal citizens; liberty is represented by both freedom of choice and the value of freedom; equality by both numerical and proportional equality of citizens.

To make matters worse, different theorists attach different weights to each of these conceptions. A utilitarian theorist of liberal democracy, let us call her Smartsinger, would want a liberal democracy to operate on the basis of the ideal, informed, well-behaved citizens' will, where liberty means getting what you value rather than getting as many choices as possible, and where each citizen's interests are satisfied in proportion to their utility function. A more libertarian advocate of liberal democracy, call him Gauthzick, would defend democratic decision-making on the basis of the ordinary citizens' will since the way people make up their minds is no one else's business. He would say that equality is numerical equality since the value of people's ideas of the good life is their own personal affair, and he would describe liberty as the range of choices people have. Anyone in between these positions will add some grains of numerical equality salt to utilitarianism, a whiff of ideal citizenship to libertarian democracy, or a teaspoon of one type of liberty to the other.

I shall try to describe minimum standards for really existing liberal democracies rather than for the ideal liberal democracy. Hence, the conditions that I introduce are in a sense common-denominators: although liberal theorists will not agree that these conditions describe their own respective ideals for liberal democracy, they will agree that they adequately describe ideals behind the practice which they want to improve on. (Cf. Walzer 1988 for the distinction between personal and social ideals.) In Sections 1.5 to 1.7 I shall (1) describe the three basic ingredients of liberal democracy, to wit, the concepts of democracy, equality and liberty; (2) describe the contradictory conceptions that exist of each element of this love triangle; and (3) formulate strict minimal conditions of liberal democracy that offer room for all these conceptions.

As suggested above, one reason why the exact meaning of the central values of liberal democracy is open to debate is that we may be trying to combine irreconcilable principles. Democracy may be short for 'the rule of the people', 'of' meaning both by and over, and it may necessarily be representative (in a direct democracy, citizens are represented on a one-to-one basis by themselves), but this is as far as agreement goes. A first cause of controversy is the degree of representation: for one school of democratic thinkers, anything other than direct democracy cannot be democracy. For others, notably the representatives of liberal democracy, the essence of democracy lies in adequately representing the will rather than the bodies of the people. But here we find a second source of conflict: what exactly is meant by the will of the people?

In an admirably lucid essay, published in 1970, J. P. Day (Day 1970) analysed the concept of the 'real will', uncovering six pairs of analytically different interpretations of it. The following distinctions are based on Day's, though the examples are mine. We distinguish:

(1a) the expressed and perceived will, as distinct from (1b) the cloaked sincere will (e.g. voting for the, in terms of issues, second-best candidate rather than the first-best who is a certain loser)

(2a) the informed will, as opposed to (2b) the un(der)informed will (voting for the Russian Liberal Democrats, knowing them not to be liberal democrats)

(3a) the immediate will, as opposed to (3b) the will for the consequences (voting for a revolutionary party, even though we do not want an actual bloody revolution)

(4a) the will for the prudent as distinct from (4b) the will for the pleasant (voting for the party that will economize, even though we know it will hurt)

(5a) the will as desire, as opposed to (5b) the will as need (continuing to drink v. kicking the habit, or voting for the spenders instead of the economizers)

(6a) the subjective will as opposed to (6b) the objective will (voting for the party that by some authoritative moral standard is definitely the wrong one).

Proponents of liberal democracy have a hard time determining their will on these alternatives. On the one hand, they believe that the persons on the Clapham omnibus should be free to make their own mistakes when voting in elections; in this respect, all twelve interpretations of the will of the people are acceptable. Yet on the other hand, the people's representatives are expected to be different: less

27

responsive to the mood of the day,[2] less susceptible also to desire and pleasure, prudent, well-informed, and certainly defiant of their private interests. What they are expected to represent is not so much the will of the people, undiluted, unpurified, as rather their *interests*, their cleansed will. And yet, paradoxically, those same representatives are chosen by and expected to represent imperfect citizens and their imperfect ideas. It is here that advocates of liberal democracy point to the checks and balances provided by control mechanisms like the division of powers, majority rule, constitutional protection for minorities, proportional representation of sex, race, class, and so forth. At the very least, if even representatives cannot be trusted to behave like ideal citizens, it is hoped that the procedures and the deliberative process in representative bodies will somehow produce a collective expression of the ideal citizens' will: sincere, informed, aware of consequences, prudent, responsive to needs and directed at a common good.[3]

A further dividing issue among liberal theorists is where democracy is appropriate. Few would go so far as to defend a total democracy, in which the democratic polity has the authority to decide on, or even actually decides on, whether women should be allowed to become Catholic priests, on whether the butcher should put the mutton or the beef on the top shelf in his fridge, on people's sleeping habits, on whom they sleep with, how long and in what fashion, what they have for breakfast, whether they put on socks first and trousers next or the other way around, at which ends they should open their eggs, how they make up their minds, make their plans and formulate their interests. Liberal theorists preach 'democratic restraint': the virtue of abstaining from interference in a private sphere even though we have the power to interfere. But again, this is as far as consensus goes. There is no agreement on when, where and most of all why democracies would be allowed to interfere in contacts between individuals. There is even no agreement on whether democracies can legitimately interfere in the individual citizens' private lives 'for their own good', at moments when they apparently cannot harm or benefit anyone else. Liberal democracies recognize a private sphere, at least in principle, one in which they promise not to interfere – but the extent of the private sphere remains open to debate.

Note, by the way, that the controversy surrounding the appropriate domain for democracy is not only one about interference in private life; its reverse side is that it also touches on the possibility of individual citizens to interfere in politics. It is one thing to extend democracy beyond elections to democratic influence after the great political decisions have been taken: to participation in policy-making and implementation, even to adjudication. All this can be construed as improving democracy as an instrument to formulate rights reflecting legitimate preferences. But it is something else to extend democracy to the agenda-setting process. Here, the sheer idea of protecting a private sphere against the will of the people may exactly become what it is supposed to prevent: questionable interference in private lives 'for their own good'.

Next, liberal theorists agree that a democracy should be an opportunity democracy: that is, citizens should have the opportunity (according to some even the duty) to participate in democratic decision-making. A government that is, say, constitutionally bound to hold elections but fails to organize them is no democratic government. Yet one of the most controversial issues in democratic theory is whether democracy should be extended beyond elections. I am not referring here to democracy after the fact or to influence before the fact as in the previous paragraph, but to the tension between representation and participation. In real life, representatives in parliaments etc. are not and cannot be the preference-reflecting machines I made them out to be. They represent parties and programmes, lobbies, interest groups, constituencies and, ultimately, individual voters and their widely diverging sets of preferences. As MPs, they are representatives; as representatives, they are parliaments in themselves. There is a tension then between their roles as representatives of the single-handedly generalized will of their constituents (or parties) and their constituents' individual wills. Hence the call for participatory and deliberative democracy, for influence of (the public debate among) individual citizens on the political negotiation process. One aspect of this issue concerns the exact institutions and fora by means of which more or less participation can be achieved, a question with which we do not need to be concerned here. The other aspect is more substantial: we may expect a participatory and deliberative democracy to change preferences through public debate and reflection, rather than to merely represent given, unexamined preferences. As said before, this book discusses the decision-making moment of liberal democracy rather than what goes on before and after, so as to focus on the principles that should guide political decision-making, not on the preferences to which these principles are applied. Nevertheless, there is reason to assume that the process of preference formation itself has an essential function in determining the possibility of a greening of liberalism (cf. most contributions to Doherty and De Geus (1996)). I shall therefore return to this subject in the last chapter.

A further controversy relates to the decision rules used in a democracy, where the choice is one between unanimity, majoritarianism (possibly outvoting minorities) and qualified majority systems, the latter in fact giving large and only large minorities veto power. There is a fair degree of agreement on the rejection of unanimity. It is reasonable to expect that many issues cannot be solved in any way without harming some people to the benefit of others, in which case a unanimity rule could easily paralyse the political system as well as society in general. But again, this is as far as consensus goes.

I mention two other sources of controversy in democratic theory only in passing. One concerns the way the citizen's will is represented, i.e. should the intensity of a citizen's preferences be counted or not? The other relates to the question of whose wills are to be represented – those of tax-paying adult men, of adults in general; of the over-30s, the over-21s, the over-16s; of individuals or regions; of foreigners, future generations, other species? Neither one of these issues is really a controversy over the democratic system as such, and in fact I do not even believe that the latter is just *one* controversy. Both are questions concerning the distribution of the right to vote. The latter issue is also one of who (or what) is thought capable of having a will at all, and one of who (or what) is thought capable of having an ideal citizen's will. These three questions concern the relative legitimacy of concrete democratic systems rather than that of the model itself, the legitimacy of which these questions already seem to presume. It is also a question of whose interests a representative body is legally obligated, respectively morally obliged, to take into consideration – and these subjects of consideration are not necessarily 'the people'. Citizens of neighbouring countries, for instance, may not be voters, yet they cannot be taken like lambs to the slaughter.

These last two questions and the one concerning the strength of preferences have little to do with the legitimacy of democracy or that of particular demo-cratic systems. They concern the duties of all rulers, not just those of representa-tives in a democracy, and they are questions of justice, of giving each his due, rather than questions of democracy, a particular way in which to give due. They are ultimately questions about the meaning of 'each', questions of equality, and will therefore be dealt with in the next section.

1.5 Equality

Equality, liberal democracy's second pillar, seems to be the most simple idea of all. One equals one, one and one equals two – apparently the most rudimentary knowledge of mathematics should suffice to understand what it means. And to some degree, that really is true. Beyond a certain point, though, controversies of medieval subtlety arise. In liberal democracies, the notion of equality performs at least two roles: one formal and one substantial. A political system assigns or dis-tributes rights on the basis of morally relevant differences and similarities between individuals and a series of moral principles for just distribution. In its formal role, equality demands that these principles are to apply equally to all individuals. In its substantial role, equality implies that the principles themselves are to treat indi-viduals as equals. The latter does not necessarily mean that individuals should *end up* as equals, nor does it say anything about the way in which people are to be seen as equal(s). All this is open to debate. In assessing the liberal democratic qualities of political systems, we do not demand that whatever a state distributes be distributed equally as measured by opportunities rather than capabilities, or by capabilities rather than preference satisfaction, desire satisfaction, market price or whatever. Given the controversies on the exact moral standard for equality, the measures actually applied are to a large extent best understood as contingent aspects of a liberal democratic system. The truly necessary traits of liberal democratic equality can be formulated in far more general terms as the numerical equality of citizens and the proportional equality of recipients. The first of these conditions is an interpretation of a central part of the Justinian definition of justice, *suum cuique tribuere*, giving each his due. As a citizen, or more precisely once recog-nized as a citizen, each inhabitant of a liberal democracy should count for one.

The usual argument for political equality is made partly in terms of liberty: in the absence of an authoritative standard by which to judge the worth of persons, no moral standard, hence no individual plan of life, hence no individual, should be privileged over any other. If there is no relevant difference between persons or their views, it follows that citizens should have an equal opportunity for political participation – one man, one vote; universal freedom of organization; universal freedom of speech – and that they should be equal before the law. In times in which we can hardly imagine that a nobleman's testimony was once worth more than that of a commoner, and that a Jew's or a woman's testimony were by definition discounted, the latter demand has become so natural that we hardly ever consider its importance.

A further criterion, proportional equality for recipients, relates to the originally Aristotelian understanding of distributive justice as treating equal cases equally and unequal cases in proportion to their inequality (Aristotle 1959, 1980). In its distributive capacities, the liberal democratic state, like any other political system, is expected to treat the recipients of its benefits and burdens according to a moral standard that meets this criterion. The exact measure of (in)equality is open to debate, but there are limits, one of which is that it should be a *reasonable* measure, i.e. that it is supported by good reasons and open to amendment should better reasons emerge. Moreover, the principle of distribution should meet the demands of *equal respect*, i.e. on the collective level no plan or conception of life is to count as better than another or should be favoured over another – prima facie. Some of the exceptions to which the prima facie condition refers are empirical: the limits posed by nature and scarcity. Others are formal: a formally equal treatment of the incompatible conceptions of the good of a Nazi and an *Untermensch* will lead, in contradiction to the intention of the equal respect rule, to the annihilation of the latter. Still other exceptions are motivated by morality, i.e. by a theory of justice: regardless of differences between conceptions of the good, individuals may be more or less deserving from an impartial point of view.

These two conditions, numerical equality of citizens and proportional equality of recipients, are relatively uncontroversial. Once we try to move beyond minimum conditions to more precise criteria for liberal democracy, we run into trouble. For one, there is no clear and precise distinction between recipients and citizens. In both capacities, people get rights that influence their chances in life. It does make some sense to distinguish between (the citizen's) influence on the formulation of distributive rules and (the recipient's) chances in life after these rules have been applied. After all, ensuring the fairness of the procedures by which distributive rules are chosen logically precedes the issue of the fairness of the rules themselves. Yet the fact remains that rights as a citizen and rights as a recipient equally determine the prospects of an individual and her equals.

Moreover, the criteria for proportional equality are – and this is an understatement – the subject of fierce debate in political philosophy. The following are only some of the topics on the agenda – completeness seems impossible.[4] There are accidental differences between humans: natural differences between individuals – in sex, constitution, condition, handicaps and qualities, but also in fortune and misfortune in the course of a life. All these are seen as arbitrarily distributed, but it does not follow that these inequalities should be somehow compensated for. Next, advantages or disadvantages gained from accidental inequality (say, wealth acquired by intelligence) are considered equally undeserved, but again, nothing in the way of redistributive obligations follows. Furthermore, to which of the following two subjects should the demand for equal respect be applied: to all plans of life *regardless* of their contents, or to all plans of life in *ignorance* of their contents? The first would demand that people get an equal number of rights, or in other terminologies an equal amount of goods and freedoms or an equal sum of natural assets and compensation. It would lead to the existence of winners and losers: some plans of life would stand a lesser chance of being successful than others. Then again, the choice of a plan of life is an individual's responsibility. The second interpretation of equal respect would give an equal chance to all imaginable plans of life, and would result in individuals having an equal share of *relevant* rights – relevant as seen in the perspective of the individual's plan of life. Thus, the hermit would get the same chance of fulfilling her plan of life as the industrialist – but it would create a world of penniless hermits and opulent industrialists. An extra source of controversy is related to this issue: should rights (or shares, or whatever) be measured in 'objective' basic acts (units, etc.) or in what they are worth to the individuals themselves? Finally, even an apparently unequivocal mathematical concept like equality has turned out to be ambiguous: which world is worse in terms of equality: one in which a has four apples and others have six, or one in which a and b have five, and others six? (For more on these medieval dilemmas, see Temkin 1986, 1993.)

As indicated in the previous section, liberal democracies know (at least) two more areas of conflict on equality other than that of the exact standard of equality: who is, or whose interests are, to be considered by the rulers, and what weight should be attached to the intensity of preferences? The distinction between recipients and citizens allows us to come some way towards an answer to the latter question. Liberal democracy would not honour the principles of equality if it gave citizens an unequal say in the formulation of distributive rules. Whether it should attach importance to intensity in distributing rights remains an open question – it depends on the answer to the third dilemma sketched above.

Finally, whose interests are to count? It seems to me that there are two minimum requirements, one of which is consideration of all human beings legitimately living within the borders of a state. Any limitation of this set seems incompatible with the egalitarian nature of liberal democracy and its perpetual quest for maximum legitimacy. Any extension of the set, on the other hand, say to include animals, might be seen to overstretch the idea of equal concern. Although extension is by no means impossible, the mark of a liberal democratic society is not so much that it is a perfect state where the ideal of equality has been fully realized, but rather – and this is the second minimum requirement – its openness to reasonable arguments for the inclusion of outsiders in the in-group. Hence, consideration of the interests of out-groups like (members of) foreign societies, for whom and in so far as a state carries a legal or moral responsibility, it is *not* a necessary trait of liberal democratic equality, unfortunate as that may be. Later on in this book I shall nevertheless argue that there are good reasons to pay attention to the interests of non-citizens and non-humans, and that a liberal democratic political system is still bound to respect these interests implicitly even if such good reasons were to be rejected explicitly.

1.6 Liberty

The third and last pillar of liberal democracy is liberty. In this context, it is probably also the most important factor: in no respect can liberal democracy and environmental concerns be so much at odds as where liberty is concerned. We can envision a concern for the environment to lead to more equality (or vice versa) in all sorts of ways: a more equal distribution of productive activities may result in a more tolerable distribution of pollution, an increased use of renewable energy sources may result in a more equal distribution of economic prosperity, better housing for the lower income classes in Northern Europe is likely to result in more efficient use of energy, and so on. We can also believe that democracy and environment are compatible: provided that a clean and viable environment is essential or even merely valuable, conditions of democracy like the demand that all available information be considered, that consequences be considered, that democratic decisions should be need-oriented rather than desire-oriented force us to pay attention to this issue. But things are different for freedom and liberty. The environment puts limits to what people can do – limits to waste production, limits to the use of resources, limits to survival. Recognizing this and translating it into rights and policies, as happens increasingly in our days with the discovery of an ever-increasing number of environmental problems (read: limits), directly implies that there will be new limits to liberty, to what people are allowed to do.

Green activists and theorists will immediately counter this and say that a more environmentally friendly society also creates new freedom: less alienation, more beauty, less haste, more room for friends, family, social contacts, in short, the valuable things in life (e.g. De Geus 1993). Whether developments like these are necessary or even likely consequences of 'greening' society and whether they are *the* valuable things in life, or even good alternatives for what may be lost, remains to be seen, but that is not the heart of the matter. What is really at stake here is a difference of opinion on the 'true' meaning of liberty or freedom: freedom as having access to the valuable things in life, or liberty as having more choices. And this may be either a result of using two distinct concepts as interchangeable (in everyday life, we do often use the terms liberty and freedom as interchangeable) or of estimating the one to be more valuable than the other.

In this section, I shall describe what I believe to be the dominant interpretation of liberty in liberal democratic thought. Part of the text (the part following the ★★★ marks) is formal, but the gist of the argument there can also be found in the informal first part. The reader who skips the formal part will therefore miss nothing essential. I shall argue that a distinction must be made between liberty and the value of liberty or distinct liberties. The latter is determined by the individual's plan of life, the former is independent. It is liberty and not the value of liberty that plays a foundational role in liberal democratic thought, partly because individual estimates of the value of choices are not a legitimate subject of collective decision-making, and partly because liberal democracy is about creating the prerequisites for individual autonomy. Hence, it is the number of options, the amount of choice that people have that matters foremost. Two conclusions follow from this. First, we must recognize that there is a genuine conflict between liberty and environment, one that cannot be reasoned away by redefining liberty in terms of the 'truly' valuable things in life. Second, if we want to reconcile liberty and environment, we should not look for solutions in redefinitions but in the discovery of legitimate reasons to limit liberty for the sake of the environment – which is what we shall do in Chapters 4 and beyond.

Over the past 40 years, the analysis of the concept of liberty has led to the existence of roughly three vested schools of thought plus a large collection of dissenters. For all, and this is a terminological point, liberty is a special type of freedom: political freedom, both as rights recognized by the polity and as the freedom required to be truly part of the polity. However, the two are often used as equivalent. Traditionally, freedom refers to states of affairs and liberty to moral rights (cf. Day 1983: 21; for 'moral', one can alternatively read 'legal'). Hence, I am free to smash the window that I was looking through a moment ago merely because I feel like it, but since it is not mine and I have neither the legal nor the moral right to destroy other people's property, I do not have the liberty to do so. If we may suppose for the moment that talk of liberties or rights to impossible actions, actions to which we are not free like turning ourselves into triangles, is meaningless (cf. Section 1.3), then liberty can be understood as a special case of freedom, and any property of freedom can, *mutatis mutandis*, be a property of liberty as well. This is certainly true in one respect for the adjective *free*, which can refer to one of three possible subjects: to things, to actions of persons, or to persons themselves (Day 1983: 17). Where *free* refers to persons, it can mean either *negative* freedom (negative liberty), the absence of coercion from others, or *positive* freedom (positive liberty) (Day 1983: 18).

In the first of our three traditions, true liberty (true freedom) is negative: freedom from interference or hindrance by others. This is the classical liberal interpretation of liberty. It has the apparent advantage of simplicity: either someone else stands in your way or no one does. Still, even a straightforward conception like this has its subtle aspects. I am obviously unfree to move if someone whose word I can trust points a loaded gun at me and threatens to fire if I move.

34

The situation becomes less clear if I do not know whether the gun is loaded or if my opponent's message is sincere. Now, I am unfree not because of the presence of an insurmountable obstacle but because my life has been put at risk. (The situations can be adapted to environmental threats.) Negative freedom can therefore exist to degrees rather than simply as a dummy variable. Second, assume that I can drive a car.[5] I am obviously free to cross a road if there is no traffic around and no law against it, and unfree if there is. Yet it seems that the notion of negative liberty is simply irrelevant when there is no road to cross or drive on – except that I may have been made unfree by others who *passively* interfered by preventing a road from being constructed, or even by never considering the construction of a road at that particular place. Thus, negative liberty covers instances of both active and passive (non-)interference. Third, it is good to note that, other things being equal, negative liberty is a kind of zero-sum game: any liberty I have against you is a limit posed on your liberty against me.

The second school in liberty theory identifies freedom with positive freedom, i.e. having the ability, the means, the opportunity etc. to do *x*. However, the meaning of positive freedom is not altogether clear; there seem to be at least two fundamentally different interpretations of it. On one view, positive freedom is an intrapersonal concept: we are unfree in this sense if we are slaves to our passions.[6] I shall refer to this view as type-I positive freedom. Steiner argued that it is a logically inconsistent understanding of freedom, since it either confuses autonomy of the self with freedom or assumes that an acting self can be unrestrained (in principle) and at the same time restrained (in fact). I am inclined to disagree with Steiner's second point: it seems to require a fully developed personality whose desires and values are clearly defined, internally consistent and nicely hierarchically ordered, but few of us are that perfect. It is not difficult to imagine having, say, a long-term inclination for vitality and a short-term inclination for vodka, and on waking up the next morning blaming *akrasia* (Aristotle's weakness of the will) for our condition. Steiner's first argument is more convincing: if freedom is mastering our passions, it is reduced to autonomy, hence to choosing a positively *valued* alternative (cf. Steiner 1983: 82). And as we shall see, there is more to liberty than the value of liberty alone – liberty and the value of liberty are distinct entities.

A more popular view understands positive freedom again interpersonally as having the necessary means to perform an action *x*.[7] I shall call this type-II positive freedom. Type-II positive freedom is a political rather than anthropological notion: it has a tremendous appeal as an instrument in critical analyses of political rhetoric. A classic example is the Marxist critique of capitalism, under which the beggar and the rich man are both (negatively) free to sleep between silk sheets and (negatively) free not to sell their labour but only one of them has the necessary means to realize these freedoms. The Kristofferson–Joplin theorem according to which 'freedom is just another word for nothing left to lose' also reflects this position. (Thanks to Caroline Wintersgill for reminding me of this example.)

The smallest third school, to which I belong myself, resists the idea that there would be a fundamental difference between negative and type-II positive freedom, and this for two reasons. First, at the hypothetical level of basic acts, both always come down to the same thing: being able to perform a basic act with one means in one way to one purpose (see Section 1.3; see also Carter 1997: 8). On this account, negative and positive (un)freedom are special cases of (sets of) such basic freedoms. Second, at the everyday level where liberty and freedom refer to complex sets of basic rights, negative and positive freedom seem to presuppose one another. A (positive) freedom to the means of survival cannot exist without a (negative) right to life, obviously, but neither can a negative right to vote be taken seriously if there never are elections, if the ballot boxes are inaccessible or hidden, if there are no pencils, paper or computers to vote with, no one to count the votes, take notice of the results or implement them.

One advantage of the last view is that it allows us to describe liberty in terms that appear to be acceptable for proponents of both type-II positive and negative liberty. The specific version of liberty that characterizes liberal democracy would then be described in terms of a combination of three ideas:

(1) the recognition of a series of purely formal rights to perform basic acts
(2) minimum interference with the formulation of purposes, and
(3) a strong commitment to empowerment: that is, to securing and enhancing the availability of the necessary means.

Liberal democratic governments promote, for instance, the availability of means like economic prosperity, education or the free flow of information. They recognize rights like consumer freedom, freedom of education, freedom of development, freedom of enterprise, freedom of expression, freedom of opinion. And they only hesitatingly restrict the exercise of these rights, particularly because they respect value pluralism and hence dislike interfering with individuals' plans of life or ideas of the good.

Now we can obviously say of any government that it recognizes certain rights, avoids some forms of interference and promotes the availability of some means. What makes a system a typically liberal democratic system is the precise content of the list of recognized, unobstructed and promoted rights. To make a long story short, typically liberal democratic rights seem to fall into two main categories: one of political rights as members of the polity and one defining a private sphere free of collective or government interference (cf. Rawls's famous two principles (Rawls 1972: 302)). In both categories, liberal democrats recognize value pluralism, or, more precisely, they recognize the existence of a practically non-reducible pluriformity of equally worthy views on the (individual's and collective) good life. In the context of this chapter, the following rights or freedoms appear to be most important:

(1) the right to vote in (free and secret) elections
(2) freedom of expression, and its counterpart

(3) free access to information
(4) freedom of trade, both as consumer and producer freedom
(5) freedom to design and pursue any reasonable plan of life (I shall refer to this from now on as 'the liberty of life'), and, within these limits
(6) freedom of lifestyle.

The first three freedoms are required, first, to facilitate democracy both in the procedural sense described in Section 1.4 and in the broader sense of deliberative democracy, i.e. the public debate in which preferences are formed, formulated, questioned and revised. Second, they are indispensable in particular for the genesis and expression of environmental preferences. Third, by making democracy possible, they also provide means to put the liberty of life (5) into practice. The latter needs no further introduction; it is what liberal thought and liberal democracy are all about, and it is also one of the reasons why liberal democracy and environmentalism are not on the best of terms. I have added (6), freedom of lifestyle, because there is a principal difference between putting a plan of life into practice and the precise way it is put into practice. Finally, freedom of trade is included because it is a necessary condition for the liberty of life on the one hand, and on the other because production, trade, business and consumerism are often perceived as the main causes of environmental problems.

These then are some of the demands liberal democracy poses on constitutional decision-making in general. Everyday real life, however, adds one important aspect: the fact that nearly every modern society is a multilingual and multicultural entity. In principle, the more this process of mixing and fusing proceeds, the greater the number of lifestyles and plans of life that become possible for individuals. In practice, on the other hand, that same combined process of diversification and unification might actually result in a cultural struggle for life and a reduction of individual choice of life-paths, if more dominant cultures or customs eliminate the weaker ones. And even though culture, like nationality, is morally undeserved (cf. several contributions to Barry and Goodin 1992), consistency would oblige a liberal democratic polity to protect value pluralism both at the individual level *and* at that of member cultures (cf. Kymlicka's (1989) conception of cultures as 'contexts of choice').

Our discussion so far of liberty in liberal democratic polities has hidden one important source of controversy from sight: namely in what respect liberty counts. Roughly (details will have to wait a moment) we can distinguish two positions in this area. From one point of view, what ultimately matters is the *number* of choices an individual has: the more choices, the more free she is. The other persuasion assesses freedom by the *value* of choices, by whether the liberties we have allow us to do the things we value. On this perspective, one meal may be worth more than ten different TV channels. Positive liberty understood as self-mastery is one among many possible sources for the value of freedom – an individual's subjective theory of the good another. It is far more difficult to bring these two conceptions together under the heading of one concept of freedom or liberty than it was to reconcile negative and type-II positive liberty.

Imagine a society where you can choose between being rich, being famous, being both rich and famous, and being a chartered accountant. You, master of your passions, choose the last. Now imagine a different society – either a dictatorship or a just and caring society where the rules prevent undesirable behaviour (cf. Carter 1995b: 6) – where your choice is limited to one alternative only: being a chartered accountant. In terms of positive liberty, there is no relevant difference between these two societies: you get what you want in both. In terms of negative freedom, or more precisely freedom of choice, there is however a far from negligible difference: rather than being free to choose the life you like, you are forced to live it. We cannot help feeling that there is some qualitative difference between the first society and the latter: freedom, here freedom of choice, matters regardless of the value attached to the choices you have (cf. Carter 1995b).

One question remains if we grant that the value of freedom matters and accept that freedom of choice and the value of freedom are distinct and not fully compatible conceptions of liberty: is freedom of choice (and implicitly negative freedom) itself a legitimate interpretation of freedom?

There are two answers to this. First, it is possible to argue that freedom of choice is a valuable interpretation of freedom from an impartial, or rather utterly disinterested, point of view: providing people with freedom of choice, regardless of the value they attach to the choices offered, shows equal concern for all individuals, plans of life and theories of the good. Thus, *even* offering people death by boiling or drowning in a greenhouse, skin cancer in the sun, radiation illness in Russia or suffocation by air pollution would be an improvement on a situation in which only pollution was available. And it is, in a way. Relative to the latter situation, people now have more room to 'make a statement', express their views on life in one final act. However, the problem with this argument is that it moves beyond the liberal *credo* that the value of choices is not a legitimate subject of collective decision-making but one of the individual's plan of life. It is indifferent to what people want, and disregards the equally legitimate view that, to individuals, the value of their liberties does matter. The former argument rests on a more fundamental point that does not have this disadvantage: freedom of choice is a necessary precondition of autonomy, it enables us to choose the maxim of our life (Dan-Cohen 1992). Without it, the whole idea of morality collapses – without choice between alternatives, there is no choice between good and bad, no responsibility, and no grounds for praise or blame. On the other hand, the more choices we have (the greater the number of basic acts which we are allowed to perform) the more room there is for autonomy.

We can now draw four conclusions, two of which have already been announced at the beginning of this section. First, redefining freedom in terms of type-I or type-II positive freedom does not invalidate the independent value of negative freedom. Concern for the environment may be translated into policy and so limit people's negative liberty. Liberty and environment are, therefore, in principle at odds with one another. Furthermore, the two cannot be reconciled by redefining them. Solutions can only be found by finding good reasons not to care for the environment in some way, or by finding good reasons to limit liberty. Unless nature has a way of adapting itself to its own or human needs'no matter what those needs are, and there is little reason for believing this, the first option is unreasonable. That only leaves the second option.

We have also seen that both negative freedom and the value of liberties are a matter of concern for liberal democrats. From this, we can draw two further conclusions. One is that liberal democracies should respect a metapreference for maximum freedom; the other that they should respect a metapreference for valuable freedom. The metapreference for maximum freedom means that individuals want more rather than fewer rights, i.e. freedom of choice, and that they want more of them by means of one of three possible instruments:

(1) distributive or stock-solutions: they want a larger share in an existing stock, even at the expense of others
(2) supply-side solutions: they want more freedom through the enlargement of an existing stock, even if natural conditions (momentarily) inhibit the creation of the appropriate rights
(3) demand-side solutions: fewer competitors for resources means less competition.

The metapreference for valuable freedom expresses the idea that individuals want freedom of choice only in so far as this is consistent with their plans of life. Hence, they want only those liberties they themselves value (for themselves or others and for their or other people's good), and they want their autonomy in this respect (i.e. the choice of a plan of life and preferences) to be respected by the polity and its citizens.

***

In this formal part of Section 1.6, I want to add a few remarks on the development of the distinction between liberty and the value of liberty, and point out that there is reason to believe that both can be quantified. Of all attempts to translate philosophical conceptions of freedom and liberty into the language of rational choice, two stand out as seminal. The first is Hillel Steiner's 'How free: computing personal liberty' (Steiner 1983), the second Prasanta Pattanaik and Yongsheng Xu's 'On ranking opportunity sets in terms of freedom of choice' (Pattanaik and Xu 1990). Steiner's theory on measuring freedom has long been neglected by rational choice authors, even though it was philosophically more sophisticated and, as it turns out, years ahead of developments in the latter field.

A fundamental axiom in Steiner's theory is the existence of *basic acts*, the atoms or measure of freedom. A basic act or action is one an individual can perform without necessarily having to perform others; it is therefore at best contingently related to other acts (Steiner 1983: 75–6; cf. Section 1.3 above). A first and quite simplistic way of measuring liberty (or freedom; Steiner uses the two as interchangeable) would be to count the numbers of acts to which an individual r has a liberty (F_r) and divide this by the total number of acts r can perform ($F_r + U_r$, r's free and unfree acts; cf. Steiner 1983: 74):

$$\frac{F_r}{F_r + U_r}$$

However, Steiner remarks, not every liberty is equally valuable – not to individual r, nor to any adjudicator of individual freedom (Steiner 1983: 80). Otherwise r, a priest who cannot drive, could be more free in an impoverished totalitarian society where religion is forbidden and traffic lights do not exist than in, say, England. Hence each basic act (x_n) has to be given a specific weight (α, β, γ etc; I deviate from Steiner's notation but the effect is the same):

$$\frac{F_r}{F_r + U_r} = \frac{\alpha x_1 + \beta x_2 + \gamma x_3 + \ldots}{(\alpha x_1 + \beta x_2 + \gamma x_3 + \ldots) + (\ldots + \chi x_{n-2} + \psi x_{n-1} + \omega x_n)}$$

But then again, measuring liberty this way could lead to rather counter-intuitive results. Let x_{n-1} be the liberty to be raped, a liberty r would value extremely negatively if it belonged to the set F_r. Now imagine that it did: ψx_{n-1} becomes part of F_r. The result would be that, despite this newly acquired liberty, r's freedom would decrease immensely (Steiner 1983: 81). In Steiner's eyes, this is a contradiction; he concludes that negatively valued liberties (let us assume: x_2 and x_{n-1}) should be ignored in the calculation of an individual's overall liberty:

$$\frac{F_r}{F_r + U_r} = \frac{\alpha x_1 + \gamma x_3 + \ldots}{(\alpha x_1 + \gamma x_3 + \ldots) + (\ldots + \chi x_{n-2} + \omega x_n)}$$

The model Pattanaik and Xu develop is initially much simpler. They assume the existence of sets X of *alternatives* and Z (the set of nonempty subsets of X) of an agent's feasible alternatives. The content of X is left unspecified: it could be bundles of goods, means to ends, Steiner's basic acts or anything else. They do not demand that liberties be individually valued, but instead assume that a measure for freedom satisfies three conditions (Pattanaik and Xu 1990: 386):

Indifference between No-Choice Situations (INS)
iff $\forall x, y \in X$, $\{x\} \sim \{y\}$

Strict Monotonicity (MS)
iff for all distinct $x, y \in X$, $\{x, y\}$ gives more freedom than $\{x\}$

and *Independence* (IND)
iff $\forall A, B \in Z$, and $\forall x \in X - (A \cup B)$, [$A$ gives at least as much freedom as B iff $A \cup \{x\}$ gives at least as much freedom as $B \cup \{x\}$]

Since they assume that in situations of no choice each alternative gives the same 'degree' of liberty, and that adding a new element to two feasible sets makes no difference for the liberty these sets offer relative to one another, an agent's total liberty can be established by simply counting the number of liberties the agent has (cf. Pattanaik and Xu 1990: 387). The authors call this a rather trivial as well as implausible result, for which they blame IND. Intuitively speaking, {train} may offers the same liberty as {blue car}, while {train, red car} can be better in terms of liberty than {blue car, red car} – but this violates IND. Hence, the authors conclude, IND should be rejected.

In his comments on Pattanaik and Xu's paper, as well as in several other texts, Amartya Sen discusses similar examples and problems (Sen 1988, 1990, 1995). He submits that a sensible analysis of liberty cannot disregard the preferences of the individuals or groups to whom liberties are assigned, since it is preferences that determine whether one liberty is actually worth more than another (Sen 1990: 470). In other words, one fundamental shortcoming of Pattanaik and Xu's model is that it does not *weigh* liberties appropriately – it merely *counts* them. Note that Sen does not dismiss the notion of freedom of choice. He merely argues that there is more to freedom than simply having as many options as possible. It is this consideration which induced Sen to present an alternative understanding of freedom as *opportunities*. We can only be free if we have the opportunity 'to live the way we would like, do the things we would choose to do, achieve the things we would prefer to achieve' (Sen 1990: 471). Put differently, freedom of choice would give an incomplete picture of liberty as we understand it. Compare individual i's opportunity set, which consists of 15 ways of being killed in a horribly painful way, with that of individual j, consisting of 15 alternative ways to be happy. In terms of freedom of choice, i and j are equally free. If j's opportunity set were limited to 14 or fewer alternatives, i would even be more free than j.

In their reply to Sen, Pattanaik and Xu (Pattanaik and Xu 1995), while admitting the sensibility of measuring opportunity freedom (read: the expected utility of liberties), observed that opportunity freedom and freedom of choice lead to different but incompatible axioms describing liberty, which in turn lead to two incompatible ways of measuring liberty (counting and weighing). The implication is that we should accept the simple fact that there are two, distinct and incompatible understandings of liberty. The same conclusion follows from a similar analysis by Gravel (1994). We have now reached the point Steiner reached in 1983.

Research has moved beyond Steiner in one respect, though: the conceptions of liberty discussed here are all insensitive to uncertainty with regard to future preferences. What I hold to be valuable today may be unimportant tomorrow if circumstances or my plan of life change. In recent years, attempts have been made by among others Puppe (1994), Arrow (1995) and Van Hees and Wissenburg (forthcoming) to close this gap. Results so far support the idea that a conception of liberty in terms of the value of liberty actually requires freedom of choice: 'keeping one's options open' appears to be the most rational course even for those who have perfected mastery of their passions. It is true that there is still research going on aimed at either reconciling freedom of choice and the value of freedom, or at reducing one to the other, or at refuting the status of one or the other as legitimate interpretations of liberty. It is also true that admitting the value of liberty as a legitimate conception creates huge problems for any polity that wants to measure its citizens' liberty and if necessary redistribute liberties – problems of interpersonal comparisons, problems relating to the uncertainty about future preferences, problems relating to the standards by which individuals freedoms can and should be valued. Nevertheless, we have little choice but to accept that both are legitimate *concerns* which are voiced, in real life, in terms of liberty. Whether or not both are valid conceptions of the one *eidos* liberty is of little importance; what matters is that both are recognized as morally *relevant* in the evaluation of any society that seeks to promote liberty.

1.7 An overview

Let me finish this chapter with a last question and, before that, a summary in the form of a list of the minimal criteria that a political institution (and its constitution) would have to meet if we would want it to deserve the epithet 'liberal democratic'. Please note that the list is not exhaustive – my sole purpose is to introduce criteria that might conflict with environmental care, and even in that respect I cannot guarantee completeness. Slightly reordered and condensed, we encountered:

(1) A series of formal criteria for the deliberation on and results of collective decision-making by representatives in the interest(s) of citizens:
   (1.1) sincerity
   (1.2) consideration of all available information
   (1.3) consideration of consequences
   (1.4) prudence
   (1.5) need-oriented rather than desire-oriented
   (1.6) directed at a common good
   (1.7) their being the product of at least an opportunity democracy
   (1.8) and at least an ordinary·majoritarian system
   (1.9) their respecting the limits posed to democracy by criteria of equality and liberty (as described under 2–5 below).
(2) Criteria for the equality of citizens:
   (2.1) political equality as electors and before the law
   (2.2) rough proportional equality for recipients, i.e. treating equal cases equally and unequal cases in proportion to their inequality
   (2.3) a reasonable measure for equality, i.e. one that is supported by good reasons, open to debate and open to amendment should better reasons emerge.
(3) Criteria for admission as 'subjects of government concern':
   (3.1) At least: consideration of all human beings legitimately living within the borders of a state
   (3.2) The possibility of extension, i.e. openness to reasonable arguments for the inclusion of outsiders in the in-group.
(4) Civil liberties:
   (4.1) the right to vote in (free and secret) elections
   (4.2) freedom of expression

(4.3) free access to information

(4.4) freedom of trade, both as consumer and producer freedom

(4.5) the 'liberty of life' to design and pursue any reasonable plan of life: no moral standard, hence no individual plan of life, hence no individual, should be privileged over any other; and, within these limits

(4.6) freedom of lifestyle.

(5) A series of criteria for the substance of liberal democratic policies:

(5.1) a duty to protect value pluralism both at the individual level *and* at that of member cultures

(5.2) a duty to show equal respect, i.e. not to favour any plan or conception of life over another

(5.3) a duty to respect the metapreference for maximum freedom of choice for individuals

(5.4) a duty to respect the metapreference for valuable freedom, i.e. for only those liberties individuals themselves value and for the individual's autonomy in shaping her preferences.

One question remains, one already posed at the beginning of this chapter: is there any reason to believe that liberal theorists, with or without these criteria of liberal democracy, will be able to see environmental problems at all? Since we still have a few chapters to go, my answer is obviously affirmative. A first set of reasons for this answer relates to some of the criteria set out in this chapter. In the introduction to this book I discussed, just as examples, half a dozen reasons why environmental issues are relevant to liberalism, liberal democracy and liberal thought. Liberal democratic decision-making is expected to meet the criteria of prudence (1.4) and consideration of consequences (1.3), and it would not be prudent or considerate to ignore these issues. Moreover, liberal democracies have duties relating to the citizens' plans of life (4.5, 4.6, 5.1–4). Hence, if people have and express environmental preferences, and practice shows they do, these almost automatically become the subject of political decision-making. At the moment such preferences go beyond demands for a cleaner neighbourhood or safe food and change into demands for reform towards a greener society, the environment and the ethical status of nature itself will come into view.

Second, even without these criteria liberal democrats should be able to perceive environmental problems once concerns are voiced – only they may not see them as environmental problems but instead interpret them in more familiar terms as problems of legitimacy, scarcity or justice. To understand environmental problems as *environmental*, liberal democrats will have to address the more fundamental social and ethical issues raised by ecologists and environmentalists. In the next two chapters, I shall introduce these issues and ask which topics liberal democrats should address if their approach to environmental problems is to count as a meaningful response to the green challenge.

2

Green political theory

2.1 Introduction

The second half of the twentieth century has seen the evolution of a most fertile family of ideas, known among others as green, environmental or ecological thought. The green family, or, to drop the metaphor, these sets of green ideas, may not yet have taken root in liberal thought but in many other respects they seem to have become omnipresent. Green ideas have permeated action in and reflection on virtually every human activity from early infancy in environmentally friendly nappies through ecojeans, biofood, even bioskunk (a biologically grown type of marihuana), ecoholidays, to death in an environmentally friendly coffin. If we neglect for a moment the reason for their emergence, these are good times for green movements and activists: the former prosper and multiply, the latter find a responsive audience for their call to arms.

However, the green family is as pluriform as it is large. The sheer amount of '-isms' used to describe green theories and people is remarkable: there is – to mention only the most important – ecologism and environmentalism, grey, green and Green, deep and shallow ecologism, anthropocentric, biocentric and ecocentric ecologism, ecofeminism and ecoterrorism. There is even a subtle difference between nature and environment. To an outsider, all these labels and distinctions must be rather confusing. This is one of the reasons why the rise of the greens has not made life any easier for social scientists, theorists and philosophers. In serious research on greens and their theories, it is by definition not enough to call everything that remotely smells of flowers indiscriminately 'green'; distinctions between positions and schools of thought have to be made. On the other hand, not all differences in shades of green are equally important. It all depends on the point of view. To the politician, the activist and the social scientist, the distinction between a deep ecocentric ecologist and a moderate anthropocentric environmentalist need not have any actual consequence since both may show the same behaviour and the same policy preferences (cf. Barry, forthcoming). To the political philosopher, the same distinction may be of more importance than any other.

We need labels for many more reasons. The more green issues have become politically and socially relevant, the more social scientists need reliable indicators to distinguish deep from shallow and Green from green. Large-scale sample research like the Eurobarometer does not allow too many or too detailed questions, hence the data it generates remain open to interpretation. The same is often true for the relevancy of indicators themselves. In research on the rise and social basis of the green movement, many a social scientist has referred to Ronald Inglehart's theory of generational value change (Inglehart 1977) in which the famous but controversial distinction between materialist and postmaterialist values plays a key role. Yet the only indicator of greenness in Inglehart's model is the value attached to 'trying to make our cities and countryside more beautiful'. It has been convincingly argued that this question, if it is an indicator of greenness at all, tells us little about greenness but more about what being green meant to Inglehart himself – in 1977 (Nas 1995). The more detailed information a social scientist needs, the more likely it is that she will design her own indicators – and possibly invent the wheel all over again.

We also need labels, a simple and concise directory or vocabulary of green ideas, if we want to investigate whether liberalism can be green. Liberals, unlike environmentalists and social scientists, *have* to invent the wheel once more. Since liberal theorists do not as yet have a theory of the environment, it will either have to be transplanted from outside, copied from the environmental discourse, or designed anew. In any case, we need a map of the terrain to find out where liberal democracy will and will not allow us to go.

As a matter of fact, such vocabularies do already exist – Andrew Vincent's article on political ecology (Vincent 1993) offering, as far as I know, the most complete listing up to now.[8] Unfortunately, directories like Vincent's have two major disadvantages. For one, they describe only those sets of green ideas actually thought to exist – either excluding other creeds through use of a by definition disputable understanding of 'green' or excluding them because the author, being human, is fallible. For another, we invariably incite disputes about the correct classification of the authors whose work is used to characterize approaches. All in all, classifications of green theories on the basis of the authors' actual writings suffer from the defect that they were never meant to be classified as authoritative statements of say ecolibertarianism or ecofeminism. They are at most seminal works in Næssism or Rolstonism or Goodinism: works in which authors expound their own views *of* concepts like ecologism and utilitarianism.

In this chapter, I want to present a more complete and less empirically contingent taxonomy, not on the basis of authors but on that of concepts. I shall try to identify the most relevant issues on which green schools of thought diverge as well as those issues that have remained hidden behind a veil of consensus. Like any taxonomy, it has its defects. Perfection or completeness is neither guaranteed nor sought for. Furthermore, it is a systematic rather than historic analysis of

46

green thought. Readers who are interested in the latter perspective should turn to other sources, such as Andrew Dobson's description of environmental justice movements (in Dobson, forthcoming) or John Barry's analysis of the history of deep ecology (in Barry, forthcoming). Last, I shall focus on issues only, not on existing schools and traditions within the green community. For a classification of that kind, I refer the reader to the taxonomies mentioned above or, even better, to the article from which this chapter is derived (Wissenburg 1997a).

In the next sections, I shall discuss green thought at the following levels of abstraction: metaphysics (Section 2.2), ethics (Section 2.3), politics (Section 2.4) and policies (Section 2.5). A full list of the concepts and theories discussed is given below. In Section 2.6, I want to make a first, rough distinction on purely formal grounds between the set of green ideas that is and the one that is not in some way compatible with liberal democracy. In subsequent chapters, the first set of ideas will be limited further by an analysis of the more substantive criteria of liberal democracy.

Three remarks are in order before we set off. First, in each of the Sections 2.2 to 2.5 and at each of the levels discussed there, I shall consider the essential ideas and topics of debate among greens in terms of *dimensions* (denoted by *numbers 1–N*) and *positions* on these dimensions (denoted by *letters a–z*). To avoid the tiresome repetition of phrases and jargon, I sometimes use these abbreviations of numbers and letters only. For heuristic reasons, the dimensions are ordered hierarchically as in a tree, assuming each dimension to deal with a question that is less deep or fundamental or abstract than the preceding one and more abstract than the following. The ordering is open to debate, but I do not see how any other arrangement could change the dimensions themselves – and it is the dimensions and positions that count here.

In describing positions I have tried to exclude eclecticism: that is, moderate or weak versions of positions like giving priority to human interests over animal interests unless special circumstances obtain. In most cases, a green theory must ultimately support one position and no other; it cannot advocate ultimately contradictory theses on the same dimension. This is not to say that in real life amendments to and mixes of positions are impossible; on the contrary. Yet for the sake of clarity we have to identify the distinguishing, fundamental, defining characteristics of theories first before we can understand modifications.

Finally, it is good to note in advance that a position on one dimension at level *A* does not always commit us to one and only one position on another dimension on any level – even if we assume that sets of ideas ought to be internally consistent. As we shall soon see, the very same ethical or metaphysical positions may for instance be combined with totally different political theories.

Table 2.1 Levels, dimensions and positions

Metaphysics	[1] (Ir)rationalism	[1a] mysticism
		[1b] rationalism
	[2] substantialism	[2a] monism
		[2b] dualism
		[2c] metaphysical pluralism
	[3] composition of nature	[3a] holism
		[3b] compartmentalism
	[4] motor of change	[4a] materialism
		[4b] idealism
	[5] causation	[5a] determinism
		[5b] voluntarism
		[5c] probabilism
	[6] 'natural' state of nature	[6a] equilibrium
		[6b] evolution
	[7] mode of change	[7a] organicism
		[7b] mechanicism
Ethics	[8] object	[8a] anthropocentrism
		[8b] intellectualism
		[8c] pathocentrism
		[8d] zoocentrism
		[8e] biocentrism
		[8f] ecocentrism
		[8g] evolutionary ethics
	[9] time span	[9a] present generations only
		[9b] future generations equally
		[9c] future generations at discount rate
	[10] nature of value	[10a] intrinsic
		[10b] external
	[11] way of attributing value	[11a] equally
		[11b] hierarchically
	[12] source of value	[12a] divine decree
		[12b] inherent to nature
		[12c] innate in humans, reason
		[12d] innate in humans, intuition
		[12e] convention
	[13] theory of action	[13a] deontology
		[13b] consequentialism
Politics	[14] scale of problems	[14a] ecological crisis
		[14b] environmental problems
		[14c] no problem
	[15] relevance of society	[15a] matters
		[15b] does not matter
	[16] relative importance	[16a] overall concern
		[16b] restricted importance
	[17] limits to growth	[17a] belief
		[17b] no belief

	[18] quantitative development	[18a] decrease
		[18b] zero growth
		[18c] limited growth
		[18d] unlimited growth
		[18e] uninhibited development
	[19] qualitative development	[19a] decrease
		[19b] zero growth
		[19c] limited growth
		[19d] unlimited growth
		[19e] uninhibited development
	[20] political theory	[20a] Liberalism
		[20b] Libertarianism
		[20c] Anarchism
		[20d] Social democracy
		[20e] Statist socialism
		[20f] Pluralistic socialism
		[20g] Conservatism
		[20h] Authoritarianism
		[20i] Fascism
		[20j] Communitarianism
		[20k] Feminism
		[20l] Political naturalism
		[20m] Political agnosticism
Policy	[21] political action	[21a] possible
		[21b] impossible
	[22] level of action	[22a] collective
		[22b] group action
		[22c] individual
	[23] radius of action	[23a] global community
		[23b] supranational
		[23c] national
		[23d] regional
		[23e] subcultural
		[23f] local
		[23g] individual
	[24] types	[24a] radical
		[24b] reformist

2.2 Metaphysics – nature and the universe

No description of green politics can be truly complete without first understanding the positions greens take, consciously or not, on questions concerning the way the universe, hence also nature, should be conceived. I call these issues metaphysical since they do not directly concern nature itself but rather our perception of it, the way we imagine the universe to be composed. To call these the most basic issues would, however, be wrong: it seems that their role in green discourse is limited most of the time to that of background premises relative to the ethical positions defended. Prior to other questions in green or indeed in any theory is an epistemological question: to what type of knowledge and argument can we appeal? I call this dimension [1] *(Ir)rationalism* and distinguish two positions on this issue: [1a] *mysticism* and [1b] *rationalism*. The choice to be made here is one between the irrational, the irrefutable, the unquestionable, the magical approach to the universe on the one hand, and on the other the rational, internally coherent or externally verifiable or falsifiable, measurable or arguable, scientific approach. There can be no arguing about the importance of this distinction in green politics: New Age greens, TM-greens, tree-huggers and other green heretics (to paraphrase Goodin 1992: 17) are involved in a totally different ball game than your everyday ecosympathetic civil servant, politician or scientist. Both parties will, at least on some decisive points, have quite diverging views on how nature operates, how humankind relates to and fits into it, on what is really happening, why it is happening, what should be done and what can be done, and most of all on what counts as a valid argument. In the remainder of this chapter, I shall follow Goodin's advice to ignore the mystic branches of the green tree, but we should note that a commitment to astrology need not prevent a green from taking a serious position on the other dimensions discussed here.

Like (ir)rationalism, the remaining metaphysical dimensions do not describe issues that are in any way typical of greens only; they are part of the heritage of metaphysics in general. One of them concerns *substantialism*: when the question arises in green thought, we either see the universe as being made up of one substance ([2a] *monism*), two, usually the Cartesian cogitation and extension ([2b] *dualism*) or more ([2c] *metaphysical pluralism*). These distinctions often play a central role in green critiques of modern Western thought and technology (cf. for instance Bosselmann 1992: 306 ff.; see also Mathews 1991). The point of this critique – at least for some greens – is that any view deviating from monism carries the risk of looking at nature in a detached, cerebral, instrumental way rather than as a substance from which we humans are part, something that is like us or even *is* us, not as something different and distinct from us.

Third, greens argue over the *composition* of nature. For some, and this links up with the issue of monism, everything in it is somehow connected to everything else ([3a] *holism*). Due to the interrelatedness of all parts of the universe, small changes in one area may have substantial consequences elsewhere, the 'where' being often unpredictable. Crudely put, the political implications of this view would be that we cannot save the otter without saving the ecosystem of which it is part, that we cannot save the latter without global change, and that we cannot do this without changing our whole attitude, culture and society. Holism and monism are two of the most distinctive features of the so-called deep ecology movement, which is inspired by the writings of (first and foremost) Arne Næss and which for a long time has been seen or saw itself as the one and only true Green philosophy (cf. Næss 1989; Bosselmann 1992: 306; Devall and Sessions 1984, 1985; Vincent

1993: 257). Other greens follow Voltaire's dictum that all consequences must have causes, all sons fathers, but that not all fathers must have sons: permanent change on a more limited scale is also thought possible ([3b] *compartmentalism*).

A further point of argument relates to the *motor of change*: does change occur in a reaction to or as a consequence of physical circumstances ([4a] *materialism*), thus making humans mere instruments of economic, ecological and social processes, or can change also originate in thought ([4b] *idealism*), thus allowing green activists to appeal to a human sense of morality or decency? One of the more practical questions in which these positions are reflected is the debate on liberal democracy. For some greens, especially those of a Marxist extraction, it is capitalism, economic liberalism, that should be seen as the cause of environmental degradation. Alternatively, it has been argued that the root of all evil lies in human greed and short-sightedness, or more generally in a disrespectful attitude towards nature, with the institutions of political liberalism operating to support rather than reform these attitudes.

Closely related to this issue is that of *causation* itself. One position claims that everything, including change, necessarily happens the way it does at the moment it does ([5a] *determinism*). There are greens who at times have taken this view rather seriously, interpreting the Gaia hypothesis as developed by James Lovelock (Lovelock 1979) to mean that nature is one and indivisible, ruled by an unchangeable invisible hand as if it were the goddess Gaia, a hand that will restore the natural equilibrium in nature any time it is disturbed (cf. the hilarious but always dependable *Green Bluffer's Guide* (Milsted 1990)). A second position holds that ideas can arise independently of the (pre)determined rules of nature ([5b] *voluntarism*). An idealist is not necessarily a voluntarist; they may well believe that our thought process follows the strictest possible rules, yet that ideas rule and shape the world. Nor is a materialist necessarily a determinist – as the existence of a third position, belief in chance ([5c] *probabilism*) illustrates.

A sixth dimension pertains to the *'natural' state* of nature: either nature is basically in a state of harmony ([6a] *equilibrium*) or it is in constant change ([6b] *evolution*). These two ideas, that of evolution and that of a natural equilibrium being disturbed by human actions, often occur in one and the same text and theory. Merging the two mostly serves either as a basis for a straightforward condemnation of humanity's catastrophic influence on the natural course of natural events (a fallacious argument confusing *is* and *ought*), or more subtly as a means of distinguishing natural from artificial change (read: the greater part of what humanity does). Note that in the latter case, references to the notion of a natural equilibrium are in fact redundant. Equilibrium and change are both valid perspectives on nature. Nature or parts of it may be in equilibrium at any given moment in time, and still evolve over time. Both are useful: they offer different perspectives on factors that influence change and stability. In short, both are respectable views. It is nevertheless essential to stress that the two – equilibrium and change – are logically mutually exclusive. If humankind is seen as part of a system that seems or ought to be in equilibrium, then whatever lasting change humankind's disturbing activities results in is by definition evolutionary, albeit perhaps not the 'right' kind of evolution. Hence the system cannot be 'naturally' in equilibrium. If, on the other hand, the system were by its very nature in a state of equilibrium, endemic development through evolution could not occur, nor could humankind be seen as part of the system. Eclectically combining the two perspectives, as happens in popular texts like Al Gore's (Gore 1992), is in fact opting for evolution. A last important metaphysical dimension is that of the *mode of change*, an issue closely related to those of composition and natural state. On one perspective, the world (the human's social and natural environment) is an organic entity that develops through continuous, long-term growth; new elements springing forth from this entity must either adapt or wither ([7a] *organicism*). On the other view, the world is seen as one gigantic machine, bits and pieces of which can abruptly be added, removed or substituted for others ([7b] *mechanicism*). Organicism is a key element in deep ecology and in related radical green theories, representing what is known as an Arcadian perspective on nature (Worster 1985). Arcadian organicistic ideas support a kind of hands-off technique of nature management: rather than trying to interfere in the processes of nature, even for the sake of nature, we should let nature follow its own course and adapt ourselves to its rules, hence living not just in harmony with nature but also in harmony with its laws. Perhaps surprisingly, the mechanicistic approach is not uncommon in Utopian green literature either. An example is Ernest Callenbach's *Ecotopia* (Callenbach 1975), which depicts the (by then more organicistic) development of the Western states of the USA after they have *radically* separated from the rest and *radically* and rapidly changed an industrial into a green society.

Finally, note that the terms environment and nature are *not* reserved for any one green school or persuasion. Although the use of the word nature is consistent with a view of humankind as part of the greater whole, implying holism, and environment with humans as isolated from their natural surroundings, implying compartmentalism (cf. Bosselmann 1992: 319), both terms are used promiscuously by authors of both persuasions.

2.3 Ethics – humans in nature

Differences in metaphysical positions are not half as important as those on ethical dimensions. It is only here that green thinking comes to have practical implications. The most fundamental topic of environmental ethical debate was and remains the *object* to which value is assigned. A policy to save the malaria virus from extinction is, for instance, not likely to be justifiable if only the object 'humankind' or 'human interest' counts. Similarly, campaigns to save endangered species stand a better chance if they defend cuddly and cute animals like seals, whales and dolphins than when they aim to protect (from the human point of view) vile, ugly or dangerous creatures like snakes and lizards. In defence of the former category, we can always appeal to the pleasure humans derive from touching them or seeing them 'play'. Lizards and snakes being less appreciated, their defence has to depend on far less popular arguments – on their moral value regardless of, or even despite, human interests, or their merely indirect contribution to the survival of cuddly species like humans.

Most positions on this issue are well-known and require little explanation. *Anthropocentrism* [8a] attributes value to humans only, or more precisely to the interests of humans. If human health can count as a human interest, then we have reason to value clean air, clear water, a healthy environment – but only in so far as *humans* need air, water, an environment. Anthropocentrism also allows us to argue, at least in principle, that the natural environment and all the creatures in it should cease to exist if that were to benefit humans: that is, if an artificial environment could better provide for human needs.

Philosophers tend to ask questions long before they become practically relevant (if ever). Once the tacit assumption that humans matter because they are humans is questioned, a deeper justification is required. Every attempt to determine why exactly one human should care about anything else but herself leads to the identification of traits that are not necessarily exclusively human. *Intellectualism* [8b] goes one step further than anthropocentrism. It attributes moral value to all beings capable of learning and understanding: in other words, it sees all intelligent beings and not only humans as capable of (experiencing) good and evil. In pre-Darwinian times, an intellectualist like Immanuel Kant designed moral laws for 'intelligent beings' rather than humans alone: he meant to include angels and God (Kant 1928; note that Kant did nonetheless argue in favour of a humane treatment of animals). Ever since Sputnik or perhaps even since Galileo Galilei, moral philosophers tend to keep an open mind with regard to the existence of intelligent life on other planets (e.g. Nozick 1974), and many (including Nozick) reckon with the possibility that primates and other higher animals may turn out to have intellectual capacities. At this point, the question where value resides loses its academic character.

Pathocentrism [8c], also known as sentientism, attributes moral value to all sentient creatures; it is most often associated with utilitarianism which, after all, sees the capacity for pleasure and pain as the measure of ethical behaviour. On this view, the realm of morally relevant creatures expands from humans via primates to borderline cases like fish – the question then being if fish suffer when they get a hook through their lips or slowly suffocate in the open air.

The next two logical steps are *zoocentrism* [8d], attributing value to all animals, and *biocentrism* [8e], attributing it to all (forms of) life. The final stage is *ecocentrism* [8f], which attributes value to all of nature (cf. among others Barry 1995a: 21; Bosselmann 1992: 251 ff.; Vincent 1993: 254 ff.). Although there are more subtle (and better) to defend these positions, these are views that can quite simply be made appealing (or discomforting) to the average, common-sense anthropocentric. The logic is easy to grasp: if we value humans even when they are severely mentally handicapped, why exclude animals? If we value them even in an irreversible coma, where is the relevant difference with any other form of life? If we waste their dead bodies by burying them rather than eating them, where is the difference between human and rock?

A less familiar position, finally, is *evolutionary ethics* [*8g*]. On this count, the attribution of value is assumed to be a matter not of one immovable, unchangeable object, but one of moral development, relative to time and place. Hence, the more we learn about the world in which we live, the more our moral sensitivity grows. We evolve from cave-dwelling insensitive barbarians to farmers who, to a degree, bond with their animals and who, even though they may suppress the feeling, recognize that animals are more than uncooked food on legs; we evolve beyond that to recognize human-independent value in inedible animals, in plants, in forests, in landscapes, in ecosystems, and so forth.

Evolutionary ethics should be distinguished from evolution*istic* ethics, 'the view that we need only to consider the tendency of 'evolution' in order to discover the direction in which we *ought* to go' (Moore 1993: 106). Evolutionistic ethics is a special, forward-looking and progress-inspired case of evolutionary ethics. The theory of evolutionary ethics is also related to another interesting perspective on environmental ethics: the so-called theory of *concentric circles of responsibility* as developed in Peter Wenz's *Environmental Justice* (1988: 310 ff.). According to this view, our individual responsibility for the consequences of our actions depends on the quality of the interaction between us and our surroundings, the interaction being more meaningful between human and human than between human and animal, plant or rock – but the latter need not be meaningless. Wenz's theory reflects what we might call the individual's moral evolution rather than that of a culture or society.

At this point, I have to confess to my own amazement that I neglected to mention a crucially important issue in the original article (Wissenburg 1997a) on which this chapter is based. It is possible to make a further distinction within each of these positions on the subject of *future generations*, although this is usually done only with regard to humans. The green critique of modern society often points to its doing harm to future generations, for example by depleting the planet's natural resources, by dumping nuclear waste in the environment or by merely using nuclear energy with the risk of causing genetic damage to the unborn. One (imaginable) position on this issue then is that future generations do not matter at all, only *present generations* do [*9a*]. For those who feel that future generations do matter, two positions remain open: all future generations matter *equally* [*9b*], or they matter at a *discount rate* [*9c*]. Some do not see any morally relevant difference between humans born in 2030 and humans born in the year 2525. If we are responsible for the welfare of non-existent beings, then it does not matter how non-existent they are (assuming the idea of degrees of existence makes sense at all). For others, there are both practical and moral reasons to argue that our responsibility either ends somewhere or diminishes over time, or both. We do not, for example, share responsibility for the starting positions in life of the next generation with anyone else, but we do share it with the *n* generations after us who all influence the fate

54

of generation $n+1$. Moreover, we may be able to predict to some degree how the next generation will want to live – with which resources and which parts of nature intact – but we have no way of predicting the preferences of the more remote future generations.

Further on in this book, I shall be rather critical of the concept of future generations and even go so far as to argue that it is of little or no practical use for us, at least not in the way it is used in academic and green literature today. The concept is popular because it is shorthand for non-existing beings whose existence is possible as individuals yet seems certain as a category – and it is easy to see why a cryptic description like this is less popular than 'future generations'. Nonetheless it obscures the fact that generations do not really exist and certainly not as successors to a place in reality – one generation inheriting the earth when the previous one ceases to exist. The thing to note at this moment is *that* the notion of future generations is popular – so popular that it has become part of nearly all definitions of sustainablity ('satisfying the needs of present generations without compromising those of future generations', see below), and sustainability in turn has become the unifying element in nearly all environmental thought.

Two other, closely related dimensions are those of the *nature* of value and of the *way of attributing* value. As to the first category, the positions are fairly well established: either we see value as something that is *intrinsic* [10a] to an object or situation, or its value is derived from *external* [10b] circumstances – again, either a situation or an object. We are talking in eschatological terms here: if in a theory value is attributed to X because of its relation to Y and so forth until an ultimate N, N having intrinsic value, then the theory must be considered as an intrinsic-value theory. If there is no N, it is an external-value theory. More down to earth, we can value animals because of what they mean to us: as food, as inedible, nasty things that are nevertheless indispensable to the food chain, as beautiful to look at, as inspiring awe, or for many other reasons – all of which ultimately make them be externally valuable, that is, not morally valuable for what they are but for what they mean to us.

Value can be attributed in either of two ways: *equally* [11a] or *hierarchically* [11b]. In the first case, assuming for example that we are dealing with an anthropocentric theory, all humans are of equal worth; in the latter, assuming a pathocentric theory, we can put a higher price on humans than on animals. An orthodox utilitarian, who recognizes no (dis)value other than pleasure and pain, would be forced into the position of pathocentric egalitarianism, that is, she would attach the same value to the pain of a hedgehog as to that of a human being.

In green theory, the following dimension is often confused with that of the object of value: the *source* of value. Humans are necessarily the mediators of value, the ones who attach value to things. Rocks do not think, they do not sense, they do not distinguish good and evil. In a way then, every theory of value is ultimately anthropogenic, yet the reasons why we accept or reject a theory of value differ.

The positions on this issue are classic and self-evident. Some see value as originating in a *divine* decree [12a], like the Bible or native American religion. Others think of it as inherent to *nature* [*12b*], as being written into the laws of nature, there waiting for us to discover them. It is also possible to think of the sources of moral value as innate in humans and detectable through *reason* [*12c*] or *intuition* [*12d*], two positions that come very close to anthropocentrism. Finally, morality may just be a mere social *convention* [*12e*].

A final distinction among greens is again a classic issue, that of the *theory of action*. The main choice here is that between two types of theories of action: *deontology* [*13a*] and *consequentialism* [*13b*]: that is, between judging human behaviour by its consistency with certain basic (rights and) duties, respectively assessing its effects. As Michael Sandel once observed, the term deontology is used in two senses: as the foundational opposite of teleology, in which case the issue is whether value is attributed to intrinsic or external origins (a distinction already covered by dimension [*10*]), and as the moral opposite of consequentialism (Sandel 1982: 3). The difference between these two dimensions is subtle but clear: in the first case we appraise objects of moral action, in the latter moral actors.

Theories of moral action, both deontological and consequentialist, are as numerous as the laws in our hearts can be; I would not want to bore the reader by listing even those few I can conceive of.[9] The taxonomy will remain incomplete in this respect, but I do not think this will in any way affect its adequacy in so far as we are concerned with the typically *green* contribution to moral and political discourse. Green theories introduce new *objects* of moral action, new criteria for the evaluation of human behaviour from the moral point of view and new weightings for criteria, but certainly no fundamental changes in the way human behaviour itself is evaluated.

2.4 Politics – the shape of a green society

Broadly speaking, green political thought has to deal with two major questions: the traditional one of the desirable shape of human society, and the new one it put on the agenda itself: how humans should relate to, administer the affairs of, nature, if at all. The answer to the latter question can but need not lead to fundamentally new answers to the former. Whether it actually does depends mainly on the positions taken on the first five of the six dimensions that will be discussed in this section. It is important to stress once more that the choice for positions on these dimensions can also be independent in a second respect. Green political ideologies are not committed to one particular view of environmental ethics or metaphysics, nor does one particular view of environmental ethics or metaphysics commit us to one particular view of green politics. Moral rules do not usually come with an instruction booklet on how to apply them in the real world with all its subtleties and contingencies (cf. Barry forthcoming). Consider the following four examples, which all seem to be perfectly coherent (though not necessarily decent) green ideologies – they share many ethical and metaphysical ideas yet deviate when it comes to implementing them:

(1) **The Green Workers Party** candidate (C) addresses a voter: 'You are being exploited!' Voter (V): 'No I ain't.' C: 'Yes you are. You could be better off under a different system in which our natural resources are not exploited, in which your air and water are not polluted, in which your children are not in danger of genetic injury or sterility!' V: 'Oh bu★★er. I see. You're right. And how do we get there?' C: 'Action. Firm action. And rigorous government intervention. That is what we need most of all now: a state that for once truly controls the economy instead of protecting the plutocrats. Capitalism will inevitably collapse because of its internal contradictions, but if we all push hard enough now, in Parliament, in the streets and in the workplace, we might just kick it over the edge a bit sooner and save ourselves a lot of trouble.' V: 'Why, that's good news, comrade. I'll vote for you then.'

(2) **The Back To Basics Party** candidate: 'You are being poisoned by your own government! You could be better off under a different system in which nature is not exploited, in which your air and water are not polluted, in which your children are not in danger of genetic injury or sterility!' V: 'Oh shoot. I see. You're right. And how do we get there?' C: 'Abolish the state! Do away with government interference, subsidies and all that nonsense. The system will inevitably collapse because of its internal contradictions, but if we abolish it now we can save ourselves a lot of trouble. Let the consumer decide for himself whether he wants to die for state-protected inert mono-polies and cartels, or if he wants to force them to become green.' V: 'You've pretty well convinced me!'

(3) **The Raving Loony Save The Dinosaur Party** candidate: 'Earth is being ravaged! By greed, selfishness, consumerism, progress, technocratic blindness, by lack of community feeling, by the immoral and unnatural acts of human-kind! Plants are being burned alive, animals massacred, tortured, whole regions of the planet are being destroyed! This whole beautiful green planet is going to the dogs!' V: 'D★mn! I'm angry! What can we do about it?' C: 'Abolish the state! Do away with government interference, subsidies and all that nonsense. The system will inevitably collapse because of its internal contradictions, but if we abolish it now we can save ourselves a lot of trouble. Let the consumer decide for herself whether she wants nature to die for second houses, second cars, second toothbrushes, or if she wants to save Creation from Sin!' V: 'My lord, you've got my vote!'

(4) **The National Green Front** candidate: 'You are under threat of extinction! Elements alien to our ecosystem are being introduced at an alarming rate by immigrants: bacteria, viruses, diseases, dangerous animals, aggressive and poison-ous plants, unhealthy habits. We should protect our ecosystem and close the borders, we should stop encouraging and subsidizing the expansion of other ecosystems, we should stop global warming and foreign aid. Each race has its own biotope – that goes for garlic and snails as well as for humans. Let us save our skins before they burn away! Vote for strong government!' V: 'Mon, you scare me. Ya're right, mon, I'll vote for ya.'

The traits that characterize green political thought are distinctly political, not metaphysical. First and foremost, a political theorist or activist interested in environmental issues has to decide on the *scale* of environmental problems. The popular (deep) green view is that we are experiencing a genuine *ecological crisis* [*14a*] threatening the survival of humankind or even that of all life on this planet. This view presumes that environmental problems are interconnected, one form of, say, pollution leading to another, partial solutions generating new environmental problems elsewhere in nature, and nearly all forms of human behaviour interfering with nature – for better or for worse. Note, however, that this position does not necessarily presume [*3a*] *holism*; it is still to some degree compatible with [*3b*] *compartmentalism*. A second position with regard to the scale of current environmental problems is that there is no crisis, there are only *environmental problems* [*14b*]. On this view, our survival or that of other sections of nature may still be at stake but need not. If it is, it is not as a consequence of a fundamental crisis caused by interconnected problems but because one or more of several, not necessarily related, environmental problems threaten one or more necessary conditions for the survival of a section of nature. Finally and at least in theory, it is possible to argue that there is really *no problem* [*14c*] at all with nature. We could expect some orthodox evolutionists to adhere to this view, not to mention some whose only interest in nature lies in its recreational functions, or some for whom the real environmental crisis is not a material but a moral crisis.

A logical sequence to the issue of the magnitude of environmental problems is that of the *relevance of society* to environmental concerns: to most greens, humans and their society *matter* [*15a*] both morally and practically. To others, the whole question of the make-up of society simply *does not matter* [*15b*]. The latter position is consistent with the ideas of a very wide range of green activists, from terrorists who, in their struggle to save nature from its most deadly enemy, humankind, would accept the elimination of individual humans, down to the respectable citizen who financially supports Greenpeace without ever linking environmental to social issues.

Green political theorists also need to assess the *relative importance of nature and society*: is their concern with the fate of nature an *overall* [*16a*] concern, limiting other interests and primary to these, or is it a matter of *restricted* [*16b*] importance, limited by other issues and secondary to them? In other words, should society be adapted to the perceived needs of nature, or is nature one, possibly minor, constraint among others on the shape of society? Very roughly, this distinction is, at least in their own eyes, the most manifest difference between Greens and greens, or deep and shallow greens, or green and grey, or true greens and hypocrites respectively fanatics and moderates. Its impact will become more clear in a moment, when we discuss *dimension* [*20*], the political theories adopted by greens.

A further familiar topic of debate concerns the *limits to growth*: either one *does* [*17a*] or *does not* [*17b*] believe in their existence. It is perhaps good to point out once more that we can be green or pro-environment, perhaps even deep green, without believing in the existence of limits to growth, if we for example believe that the real environmental crisis is a moral crisis. Moreover, a simple and formal recognition of the limits to growth commits us to neither an opinion on how close we are to those limits (cf. dimension [14]), nor to one on how important those limits may be (cf. dimension [3]), nor, consequently, to one on the seriousness of environmental problems. We can believe in material, economic or social limits to growth without immediately interpreting these as environmental problems – it takes more to be green. These are all analytically distinct issues.

Next, greens can take sides on the issues of the *quantitative* and *qualitative development* of the human population – meaning, less formally, population growth and economic growth. Possible positions here are: *decrease* [18a, 19a] of population, respectively a shrinking economy; *zero growth* [18b, 19b]; *limited growth* [18c, 19c], also known as sustainable growth; *unlimited growth* [18d, 19d], a possibility both for those who do not believe in limits to growth and for those who believe that nature itself will restore whatever balance might be upset by too high a rate of development or too hazardous a type of development; and lastly *uninhibited development* [18e, 19e], meaning abstaining from any population policy or economic policy rather than actively promoting (either decrease or) growth. Note that the ambiguous phrases *sustainability* and *sustainable development* are used as euphemisms for both sustainable and zero growth.

Finally, greens can be distinguished by the type of *political theory* to which they adhere: that is, the type of *human* society they advocate. In contrast to most dimensions discussed so far, the positions on this one are not necessarily mutually exclusive. They can occur in all sorts of combinations or mixtures. A clear example is feminism, which, like ecologism, can easily be combined with many other political theories. Nor would I want to pretend that the list I am about to give is complete (it surely is not), or that it is new. With two exceptions, the positions listed describe very old or very familiar political theories. The point to note is that, in principle, greens can support any political theory, any form of political and social order. Their green concerns, whether overall or restricted, do not have to result in totally new blueprints of perfect or imperfect societies; they can also lead them to put new (green) constraints on old theories, find new arguments for old theories or against alternative theories. Hence, a green theorist can also be, say, a liberal. Depending on whether her concern with nature is overall or restricted (cf. dimension [16]), I would describe her as an ecoliberal respectively a liberal environmentalist.

Andrew Dobson has argued that not all ideologies are equally compatible with ecological concerns, particularly not if the latter dominate other political concerns (cf. position [16a], or Dobson's description of ecologism (Dobson 1990: 13, 1995: 3 ff.)). I believe he is correct in this: not every form of human society is compatible with every view on human-nature relations. I have no intention of denying this when I say that greens can, at least *in principle*, support any other political theory. The latter may have to be adapted to deal with inconsistencies, and adaptation may turn out to be impossible, but there is no reason to assume beforehand that one set of prescriptions on interhuman relations and another on

59

(primarily) the relations between human and nature must necessarily be incompatible (cf. the four examples at the beginning of this section). The dimensions I introduce in this text are primarily meant to describe possible green theories as a first necessary step towards answering further questions, like whether these possible theories can also be logically valid and empirically supported. The list, for which I have borrowed at least [20e] and [20f] from Vincent (1993), looks as follows:

[20a] *Liberalism*
[20b] *Libertarianism*
[20c] *Anarchism*
[20d] *Social democracy*
[20e] *Statist socialism*
[20f] *Pluralistic socialism*
[20g] *Conservatism*
[20h] *Authoritarianism*
[20i] *Fascism*
[20j] *Communitarianism*
[20k] *Feminism*
[20l] *Political naturalism*
[20m] *Political agnosticism*

The last two entries on this list require some further explanation. Political agnosticism is included to cover cases in which a theory or theorist avoids all mention of a favourite or preferable political order. Political naturalism refers to the hypothetical category of genuinely new and typically deep green 'back to nature' ideologies. What would distinguish these political theories from other entries on the list is that they would want society to be rearranged *as* a part of nature following the perceived rules of nature. We could imagine here, for instance, an eclectic theory dividing human society into bio-, morpho- and georegional societies, with cultures, customs, economies and consumption patterns conforming to the ecological structure of each particular region. To some extent, Callenbach's earlier mentioned *Ecotopia* (1975) satisfies these conditions.

Although I promised to keep this taxonomy purely analytical and avoid all discussion of really existing schools of green thought, there are two groups of terms used to designate green philosophies that really must be discussed here because I often use them and because they are ambiguous. One of these is a group of three pairs of opposed notions: deep and shallow green, green and Green, and green and grey. As a rule we can say that *grey* is used to denote every idea that is consistent with every denial of an environmental crisis [14bc]. The use of the term green is far more indefinite; there seem to be at least three meanings. First, it is universally used in its most imprecise and general meaning as 'everything that remotely relates to a political interest in the environment', as I for instance have done throughout this text. Second, *green* can be used in a *broad* sense as the opposite of grey, to denote every position consistent with [14a], the recognition of a global environmental crisis. Finally, green (lower case g) is used in a *strict* sense as the opposite of *Green* (capital G). In the last case, both are consistent with [*14a*] but Green refers to theories in which nature is of overall importance [*16a*], whereas in green theories (lower case g) the interest in nature is one among others [*16b*]. The labels *deep* and *shallow* are used in two ways: either as synonyms of Green and green, or to denote consistency with the metaphysical and ethical ideas of deep and shallow ecology (see previous sections). In real life, the difference between these two interpretations is negligible, Green theories usually being consistent with deep ecology and green ones with shallow ecology, but as argued before, usually and necessarily are not the same thing. A shallow ecologist is not predestined to be lower-case-green.

The second group of terms, *environmentalism* and *ecologism* a.k.a. political ecologism, are often used to denote anything green and a particular brand of green. They are, like deep and shallow, also used as synonyms of green respectively Green. Finally, they can also be used in a more determinate way. Both views not only presume the existence of an environmental crisis and the limited (environmentalism) respectively overall (ecologism) importance of the fate of nature, they are also explicitly political creeds, addressing environmental as well as political and social issues. Moreover, both presume the existence of limits to growth, ecologism opting for either decreasing or zero growth, environmentalism for limited ('sustainable') growth. However, neither one is necessarily committed to the idea that politics and politicians can or should influence growth.

Throughout this book, I shall keep on using the terms green and environmentalist in the broadest possible sense: that is, to include all shades of green. I do not do this in order to silently change the question 'can there be a green liberalism?' into 'how green can liberalism be?' – it is simply a matter of convenience. Besides, the answer to the latter question may well be zero, although I hope to prove that it is not.

2.5 Policies

In comparison to the previous three levels of debate, characterizing green policy theories is a very simple and straightforward affair. The first question for greens here is whether they believe *political action* to remedy or thwart environmental problems is *possible* [*21a*] at all or *not* [*21b*]. Judgement on this issue depends mainly on the positions taken on a combination of the dimensions mentioned earlier, especially materialism/idealism [*4*], causation [*5*] and equilibrium/evolution [*6*], but also on quite practical considerations like the availability of means and resources, of political support and legitimacy.

Second, the green policy-maker has to decide on the *level of action*. One possibility for green policies is that they are *collective* [*22a*]: that is, deliberate attempts by the institutions of a society to transform or preserve a *status quo*. Most environmental policy is of this type, and most political action is directed at promoting or inducing such collective policies at the national, international or subnational level. A second option is that of *group action* [*22b*], in which one or more segments of society neglect the collective route and start operating on their own. Famous examples of this approach are Greenpeace, the RSPCA in its early days and Earth First! If neither of the above applies, green policy will be a purely *individual* [*22c*] action. The latter is especially an option for those who believe in invisible hands, the self-regulating capacities of the free market (green consumerism) or in the ineffectivity of political coercion. Alternatively, we may simply believe there is a need to set an example; this seems to have been the Unabomber's source of inspiration. (The Unabomber, by the way, presented himself as a group; cf. Unabomber 1995.)

The issue of the level of action goes hand in hand with that of the *radius of action*. The positions on this dimension hardly need explanation: green policies can be directed at the *global* [*23a*] community, a *supranational* [*23b*], *national* [*23c*], *regional* [*23d*], *subcultural* [*23e*], *local* [*23f*] or *individual* [*23g*] level. Decisive for any choice in this field is not whether the problem at hand concerns the whole planet, a biotope or merely the tree on the corner of the street – but where, at which tier, the necessary means and support can or should be mobilized. The range of the problem does play a role in this decision, but so do considerations of a political, ethical and metaphysical nature.

Finally, green policies can be basically of only two *types*: *radical [24a]* or *reformist* [*24b*]. Here, I borrow the famous and elegant distinction Karl Popper made between Utopian and piecemeal engineering (Popper 1960: 64). The two describe fundamentally different approaches to problem-solving. Piecemeal engineering presupposes that problems should *in principle* be solved separately, one by one, one after the other, in their smallest possible shape, with the lowest possible amount of energy at the lowest possible speed. The idea behind this is that a policy aimed at realizing X aims to realize X and not Y. Now according to Popper's falsificationist view, we can never be sure that we have complete information, nor that if we did (though on this view truth is a red herring), we could have complete control over the unlimited amount of information that 'complete information' would require. Since this is so, any change can have unexpected consequences, that is, result in Y rather than X, which makes it advisable to design policies in such a way that they cause the least possible amount of expected damage and offer a maximum opportunity for testing and adapting them. Utopian engineering, on the other hand, presumes that problems should be solved as radically, speedily, completely, directly and fundamentally as possible, precisely because there is always a chance that something (pressure groups, new issues and problems, second thoughts) may come in between the first steps towards X and the realization of X. Though these two approaches clearly differ at a fundamental level, it is equally clear that, under the right circumstances, the two can result in exactly the same actual policy. They need not, though.

In line with the distinction between Utopian and piecemeal engineering, we can make a rough distinction between two popular approaches to green policy.

One is *ecological modernization*, the background of which has been described in detail by among others Albert Weale (Weale 1992). The picture that emerges from this description is that of a typically active [21a] and collective approach [22a], applicable at any level [23a–g] and modest in its aims and expectations [24b]. It can be contrasted with what might be called *ecological Utopianism*, the preferred approach of radical ecologism: also active [21a] and collective [22a], also applicable to every tier in world politics [23a–g] and probably applied to all at once, but distinguishing itself from ecological modernization by its radical methods and ambitions [24a].

2.6 Compatibility

So far, I have discussed 24 dimensions on which green theories of society, of politics, of policy and of moral behaviour can take positions. Even excluding irrationalism, this leaves us with a staggering 39,731,628,000 'possible' theories to choose from, each of these being green in the sense that they are 'remotely related to a political interest in the environment'. Although I did occasionally exclude possible green theories (that is, combinations of positions) that were clearly logically inconsistent, in particular any combination of [16b], attaching limited importance to the fate of nature, and [20l], political naturalism, I did not consider it my principal task to restrict the list to *consistent* or empirically supported green theories only. Whether or not a green theory is logically consistent depends on the compatibility of the positions, which in turn depends on their exact formulation, and the latter issue is obviously open to philosophical debate.

Since virtually no limits have been put on the characterisation of theories as green in the widest possible sense, it might seem that anything goes, that there is no way of characterizing a theory as *truly* green. This would be a mistake. The set of possible green theories, of possibly truly green theories, can be limited in at least two ways. First, there is convention. As indicated above, there are certain conventions restricting the use of the term green to theories that recognize the existence of a global environmental crisis [14a] and in some cases restrict it even further to distinguish between those that are not–really–green (lower-case-green theories in which the interest in nature is but one among several others [16a]) and those that are capital-G-Green theories in which the importance of nature overrules every other interest [16b]. Yet we should remember that these are *merely* conventions.

Second, there is the more promising alternative of consistency analysis. It seems elementary to demand that a possible (or rather, in this context, possibly) green theory be consistent in two respects: (i) with regard to known facts and (ii) with regard to the internal structure of the theory. I have already said something about the latter issue, logical consistency, above; suffice it to say here that though it may not be a simple demand, it is also one we cannot do without. It is a necessary requirement for any theory if it is to be taken seriously: specifically, if it is to be taken seriously as a *green* theory.

The term 'known facts' in the first condition should not be interpreted too strictly. If there is such a thing as a global environmental crisis, then any theory that does not take this fact (in the strict sense of an empirical fact) into consideration cannot truly be 'interested in the environment'; it cannot be called green. But similarly, if it were true that humankind should not intrinsically value life as such but only some forms of life, then any theory inconsistent with this fact (in the related sense of an ethical fact) cannot be green either. We must admit that unearthing a fact, discovering that something actually is or is not the case, whether ethically or empirically, is not easy, to say the least. Yet we have to assume its possibility as a second necessary condition for judging if theories do or do not make sense as *green* theories.

In this final section I want to focus on the second demand for consistency, i.e. the internal logic of a green theory. I want to determine which green ideas are, on a purely formal reading, compatible with liberalism and hence within which limits we may expect a 'green' liberalism to be possible. In other words, I shall mark off an area within the set of all possible green theories, using the criteria developed in Chapter 1, and call this the set of possible theories of green liberalism. The positions thus left open to green liberalism will be rephrased in Chapter 3 in terms of an agenda of ordinary, everyday political questions relating to environmental problems to which we shall seek a liberal democratic answer in later chapters. At the end of the book, when we have our answers, we will also be able to say how *truly* green green liberalism can and should be.

The consistency analysis could start at any randomly chosen point within the previous four sections and move from any issue to any other issue anywhere else in the text. To bring some order into our discussion, I shall start with environmental policy issues and then move on to politics, ethics and metaphysics.

It is clear that a liberal democratic polity cannot use, promote or accede to radical individualistic methods (read: ecoterrorism) for any purpose, ecological purposes included. What is perhaps less clear is that it cannot be reconciled with ecological Utopianism either. Consider some of the ideas with which ecological Utopianism (as a policy approach) is associated: it conceives of the environmental crisis as a principally moral crisis; the discovery of the natural 'limits to growth' merely highlights the need to reconsider humankind's relation with its natural surroundings. Ecological Utopianism proposes both moral and social reforms, i.e. the typically Utopian combination of a compulsory restructuring of human preferences and desires and a revolution in the structure of society. Proponents of ecological Utopianism often adhere to a distinction between an (ethically preferable) ecocentric and a (morally perverse) anthropocentric morality, and a critique of the one-sided development of Enlightenment rationality towards technical rationality and dominance. Ecological Utopianism is not one single and united green theory, though. Its advocates range from Utopians in the strict sense like

Bookchin (with sometimes little interest in purely ecological issues) to ecomoralists like Næss. Nor are its themes necessarily expressed in the traditional form of Utopian proposals. What unites them are two things: having radical ideas and being radical in their radicalism. It is this second trait that makes them impossible bedfellows for liberals.

Assume for a moment a worst-case scenario: at a given point in time, a polity has full authority in all political matters including the environment; there is an environmental crisis of the first magnitude, say a sudden and complete breakdown of the ozone layer; there is one and only one remedy; it requires immediate implementation; it is also a remedy that will disrupt each citizen's life prospects since it requires a return to Neanderthal life; and it is both perceived as being in the citizens' best interest and universally accepted, on every account of the will, by the citizens themselves. We should now note two things. First, this could still be an instance of the reformist approach to policy-making because it comes down to solving the smallest possible number of problems at the lowest possible speed with the least possible energy. Second, if it were an instance of the radical approach, it would presuppose that all debate on criteria of equality and concern should be terminated *as a matter of principle*, and that civil liberties should give way to the overwhelming importance of the citizens' best interest, again *as a matter of principle*, the principle being that what needs to be done needs to be done as quickly as possible before anyone changes their mind.

As a matter of fact, it would seem that the only policy strategy that is really compatible with liberal democratic principles is ecological modernization. Ecological modernization conceives of the environmental problem as a matter of fact rather than morals. It is a means-oriented approach, built around the assumption that *given* preferences and desires ought to be and can be answered in a different, ecologically responsible way. It does not neglect the processes of preference-formation in civil society but assumes preferences to be inviolable. It presumes that it is not the task of the political system to adapt people to an ideal society but the other way around.

Other, non-collective, reformist strategies like green consumerism or cartel agreements are essentially control mechanisms of a civil society or, in a broad sense of the term, a free market. They give us the usual problems involved in combining economic liberalism (the free market) with political liberalism (liberal democracy). For one, the free market, even if it operates in a pro-environment way, dismisses with all democratic debate on the criteria for (in)equality since it recognizes only one such criterion: financial power. It may also violate the individuals' and cultures' freedoms of lifestyle and life-plan, and in the case of cartel agreements particularly consumer freedom, simply by making some goods unavailable. The only way to fit these strategies into a liberal democratic polity would be by making them operate within possibly very broad but certainly strictly defined borders set out by a policy that can deliberately accommodate liberal democratic demands, i.e. one of ecological modernization.

However, first sight can be deceptive. Whether these non-interventionist stategies are actually incompatible with liberal democracy ultimately depends on who should count as a 'subject of consideration'. One answer to that question might just be – purely hypothetically speaking – a version of the deep ecologist's egalitarian ecocentrism in which some human interests can be sacrificed for non-human others. To settle this issue we will therefore first have to consider (as we shall in a moment) which environmental ethic can complement liberal democracy and which of them fits which green policy.

Before that, let us consider how liberal democracy scores on the level of green politics. The first thing to note here is that liberalism is (in principle) compatible with every possible position on the dimension of the scale of environmental problems [14]. If liberals do not see any environmental problems, there is no problem of compatibility whatsoever. If they do, and we have argued that they should, then the question of compatibility becomes a practical and not a theoretical one. The answer would depend on the nature and depth of the ecological crisis or environmental problems and on whether or not a liberal democratic system can, in practice, deal with them.

On the issue of the relevance of society [15], liberal democracy is bound to take a far more definite position. Since it is, traditionally, a system for which the desires and needs of humans are its *raison d'être*, it must consider society relevant and in addressing environmental issues at least take humans as its point of departure. Green liberalism cannot be a theory of the healthy environment if it is not first of all a theory of the healthy society.

Although liberal democracy *may* find environmental problems to be of restricted importance (dimension [16]) relative to other matters – say, liberty, the quality of liberty, basic human needs – it is not predestined to. Nature puts side-constraints on the just society: that is, it pre-empts possibilities and choices both for individuals and for the system as a whole. Nature in all its guises is a necessary precondition for human existence and therefore also for the existence and survival of liberal democracy. Thus, even though it may set out as an anthropocentric political theory, and even though it may see individual environmental problems as less significant or acute, it cannot afford to see the category environment itself as a really minor issue. As to the limits to growth thesis [17], there is no reason to assume beforehand that liberal democracy is incompatible with a belief in either infinite or finite growth of production, consumption, population. Liberal democracy as described here is a theory about the regulation and harmonization of individual preferences and about protecting the procedures through which preferences can be formed. Even though it cannot meet impossible desires, it does not impose an *a priori* criterion on what these preferences should be. In line with the conclusion drawn in the last paragraph, the only thing that can be said with certainty is that if the limits to growth thesis rest on sound empirical data, liberalism cannot afford to ignore it. In Chapters 6–8, I shall investigate this possibility and further determine where liberal democracy stands on the issues of the quantitative [17] and qualitative [18] development of society.

Finally, we must at this point ignore the question of the compatibility of the principles of liberal democracy with specific political theories [20]. We are, after all, primarily interested in whether liberal democracy and environmental concerns are compatible. The latter may or may not require a restructuring of political systems along the lines of political naturalism [20l], and this particular compatibility question will therefore have to be answered in later chapters. Whether liberal democracy is compatible with other political theories is a question that has first of all been discussed in great detail elsewhere in political philosophy, and one that is moreover of secondary importance in this context.

Although no position in environmental ethics necessarily commits us to a specific type of politics or policy, nor vice versa, some combinations are more compatible than others. For the sake of brevity – these are issues with which we shall deal in far greater detail later – I shall assume without argument that liberal democracy is compatible with external (instrumental) value, as it obviously already is, as well as with intrinsic value [9]; the latter does not exclude the former. I make the same assumption with regard to theories of action [13] and future generations [9]. I also assume without (much) argument that liberal democracy accepts only human sources of value [12], not divine decrees or nature itself. Only the first two can accept the need for human recognition of values, i.e. they take preferences and theories of the good seriously. There is no reason why individuals cannot or should not be allowed to believe in divine decrees or in values as 'naturally' revealing themselves to us in the world, but making these the cornerstone of a liberal democratic polity would contradict its commitment to impartiality.

Furthermore, and for the sake of simplicity (subtlety can wait until later chapters), I reduce the number of possible objects of environmental ethics [8] and ways of attributing value [11] to six combinations: the hierarchical and egalitarian versions of anthropocentrism, pathocentrism and ecocentrism. The latter threesome are both the major schools in current environmental ethics and, in their relation to environmental policies, representative of the problems other schools (*mutatis mutandis*) would have with liberal democracy.

For starters, liberal democracy and an anthropocentric hierarchical ethic are clearly incompatible. The latter violates all principles of equality as well as the criteria for concern: it would make some individuals outsiders relative to more worthy others. With one exception, the same is true for pathocentric and ecocentric hierarchical ethics. Only if humans are considered as mutually equal in worth and if humankind stands at the top of the ladder of concern can these types of environmental ethics be successfully blended with liberal democracy.

Eco- and pathocentric egalitarianism escape these objections: there is no reason to believe that either one must necessarily reject proportional equality, deny political equality to humans or go for the absurd option of enlarging the electorate with rabbits and cabbages. In fact, both are pleas for the inclusion of other interests in political deliberation, for 'extending the circle of compassion', as it has been called.

Anthropocentric egalitarianism and liberal democracy are perfectly compatible: the former has always been the dominant morality within liberal democratic thought. There is one area in which it seems doubtful if any environmental ethic other than anthropocentric egalitarianism can be reconciled with liberal democratic ideas: that of civil liberties, in particular those liberties that concern the liberty of life. Clearly, any material or mental sacrifice made by x or any possible benefit denied to x on behalf of y can be, and can be understood as, a restriction of x's liberty (cf. Section 1.3). Yet even if the 'can' in this proposition were (incorrectly) replaced by 'will necessarily', the y in it could be a fellow citizen as well as a cuddly species or a foreign landscape; the difference is only gradual. The interests of any newcomer y – human or non-human – can legitimately limit an individual's choices and opportunities. The addition of more subjects and interests to a polity is therefore nothing new, nor is it a reason why liberal democracy would have any specific problem with an extension of the circle of compassion. It may have *bigger* problems – we shall come to that in a moment – but it is not an essentially *new* challenge.

A more consequential reason to suspect that liberal democracy and non-anthropocentrism might conflict is that the sacrifices made or benefits foregone 'on behalf of y', y being something non-human, do not advance or protect y's plan of life. The best if not only justification liberal democracy has for limiting one individual's liberty of life is that this will in some reasonable way serve to protect the same liberty of another or others. With the possible exception of some animals, it seems, however, that no non-human entity has anything remotely similar to a plan of life, nor a capacity to use the liberties of life, nor therefore anything that can serve to justify limits to human liberty.

At this point, we should avoid getting entangled in the famous and incessant debate among green philosophers on criteria for the moral relevancy of subjects. In this chapter, I have opted for the idea that what matters about people relates to their having or executing plans of life, but there are alternatives. However, I shall argue later (Section 4.4) that these alternatives either do not make a difference or, if they do, would alter our idea of liberal democracy beyond recognition.

For now, the only conclusion we can draw is that the question of the compatibility of liberal democracy with types of environmental ethics other than anthropocentric egalitarianism can only be answered affirmatively if one of two propositions is true. Either there must be indirect reasons for attributing value to non-human nature, reasons that make these parts of nature instrumental in advancing the human liberty of life, or liberal democracy must offer room for the idea of intrinsic value. We shall discuss this issue (the nature of value, dimension [10]) at length in Chapter 4.

In the area of metaphysics, liberal democracy excludes only one possible position: mysticism. The way nature is seen to be composed will limit the options open for environmental policies – holism, for instance, will demand a more cautious version of ecological modernization than compartmentalism – but it will not change, invalidate or disprove the moral principles of liberal democracy. The same observation applies, *mutatis mutandis*, to other dimensions. Liberal democracy excludes mysticism, though: in a universe that is conceived of as unquestionable, irrefutable, unmeasurable, unpredictable and in general beyond human reason, any attempt at realizing, implementing, even inspiring society with, liberal democratic values will be considered futile.

Pending the results of a more thorough analysis in later chapters, the results of my initial probe can be summarized as follows. Assuming that a polity should be liberal democratic, its attention for environmental issues will have to be inspired by an egalitarian appraisal of the value of the environment, probably by anthropocentric egalitarianism. It cannot be hierarchical unless it is still egalitarian for humans. Its concrete policies may offer room for the non-radical private initiative, but only within limits set by an overall strategy of ecological modernization. Being liberal democratic, the polity would offer room for dissenting opinions on environmental ethics. Yet it cannot implement them except on the basis of an anthropocentric consideration, i.e. when the means and ends of dissenting views on environmental politics happen to coincide with the interests of its citizens. What all this means is that we have discovered a window of opportunity: there is no prima facie reason why liberalism could not be green. What needs to be done next is to establish why liberalism has never been perceived as green – and how this can be mended.

3

The possibility of green liberalism

3.1 Introduction

In the previous two chapters, I described the ideas that define liberal democracy and the set of ideas on which green theories, theories of nature, the environment, sustainability and ecology, are based. We have seen that the two are, to a large degree, prima facie compatible. I have already discussed some practical, real-world reasons to explain why there is a need to enrich liberal democratic thought with a green dimension. The main question that will be dealt with in this chapter is in which precise areas, on which issues, liberalism should be so enriched, and for which reasons.

Over the last 20-odd years, a set of issues collectively labelled as 'the problem of sustainability' or 'the environmental problem' has found its way into the public debate and private life, into economy, politics, science and philosophy. It has become one of the most serious challenges for modern liberal-democratic societies, as well as a widely discussed theme in several scientific and philosophical disciplines. Recently, this resulted in the first attempts to reconcile green concerns with existing democratic institutions (e.g. Doherty and De Geus 1996), focusing on the kinds of conceptions of democracy Greens (should) adhere to. Curiously enough, though, the environmental issue has only had a marginal influence on developments in one extremely important area of research: the mainstream paradigm in (moral and) political philosophy. This so-called liberal approach to political theory is concerned with the moral foundations of liberal democratic societies and, in turn, informs the practices of politicians, parties, policy-makers and voluntary organizations within these societies. The mainstream debate has mostly followed its own agenda, which is defined by the theoretical relevancy of subjects rather than by their practical urgency.

Theorists in this field are virtually silent on purely ecological issues, and in particular on the issue of nature's intrinsic value (as opposed to external, instrumental or user value, value as a resource for human action; cf. Goodin 1992: 19 ff.). The ecological exploits of mainstream theorists are, as we shall see, limited to more or less incidental discussions, often literally in footnotes, of related issues like the just distribution of the world's natural resources, obligations towards future generations and those towards animals.

In the first part of this chapter I shall be concerned with three descriptive questions, the first of which is how liberals (until recently) understood that aspect of 'nature' which we now call 'environment'. I distinguish four phases: classical and 'sensualist' liberalism (Section 3.2) until the 1960s, modern-day Rawlsian liberalism between roughly 1960 and 1980, and latter-day 'greening' liberalism (Section 3.3). Section 3.3 also deals with my second question: what has changed in liberalism since the environmental crisis was invented? Although its answers are surprisingly radical, liberalism still deals with mostly marginal questions. I conclude that theorists in this field, with only one notable exception, are virtually silent on purely ecological issues, in particular on the issue of nature's intrinsic value. Third, in Section 3.4, I have two goals:

(1) to sift out and list those ecological issues that are a real but neglected challenge to the liberalism of the future; and
(2) to indicate the directions in which the liberal debate should develop if it wants to address such ecological issues adequately.

Most of the topics I introduce here already rank high on the agenda of the competing ideology, ecologism – in particular the intrinsic value of nature, the representation of non-human interests, population growth and ecological reasons for limits to property rights. Yet the green agenda contains more than moral and theoretical problems. Basically, what unites all environmental theorists and activists from greyish green to deep Green is the very down-to-earth question of survival. Many issues on the green agenda relate to the identification of the causes of environmental problems (or, if you like, the ecological crisis) and the description of solution strategies. But questions like these cannot be answered, cannot even be raised, without the existence of green issue number one: what is actually going on? What exactly are the practical environmental problems that threaten our and nature's survival? A green liberalism that can give abstract answers to questions of means and ends is purely speculative if it simply *has* no problems to deal with. In Section 3.5, I therefore give a brief review of environmental problems as such and select a few of them for further consideration caused by the unusual political problems they may pose for green liberalism.

3.2 Classical and sensualist liberalism

Political philosophers are, by definition, interested in the *polis*, in the relation between government (court, proto-state, state) and civil society – and as an inevitable consequence also in humans, their mutual relations and their desires, needs, wants and preferences. What we today call 'the environment' plays virtually the same role in (classical) political philosophy as 'the stars above' or 'the existence of extension and cogitation' or 'the digestive system/the body': it is one of the areas in which the Laws of Nature apply, laws that limit our (human) freedom to do as we like. The one role it does not play is that of a politically interesting subject, of something that somehow presents political problems or inspires some kind of policy.

The above is not only true for political philosophy in general but also for liberal political philosophy in particular. Before I go on to describe the classical liberal ideas of nature in more detail, let me first indicate what I would designate as a liberal political philosophy. It is clear that any such demarcation of a territory is done with the benefit of hindsight. John Locke, for example, did not describe himself as a liberal and hardly used the word liberal – at least not in the modern political sense of the word. Classical liberalism exists because modern liberalism does: classical liberalism is what modern liberals embrace as their intellectual ancestry.

Now modern liberalism itself is not typically one school of thought. Active American libertarians are, by European standards, liberals *pur sang*, yet they themselves use the term liberal to refer to – by American standards – leftish interventionists or, even more insultingly, as a synonym of social democracy. The former group would, both by European standards and by American academics (cf. Kymlicka 1990), in turn be described as liberal egalitarians. Finally, modern European social democrats also claim both to have a liberal streak and a partly liberal ancestry.

I shall not argue here about what would be the most or true liberal school either in politics or in political philosophy, but will instead adopt a liberal attitude and identify the family of 'typically' liberal ideas as that which defines liberalism; it is in the nature of a family concept to allow wide variations on a theme and not demand full devotion to any one particular note. Nor shall I bother to design a complete list of all the elements of the family liberalism. The following three criteria, a condensed version of the criteria for liberal democracy, are sufficiently uncontested to serve as a guide to identifying both classical and more modern versions of liberalism:

(1) Equality of citizens before the law and in politics
(2) Proportional equality of treatment by governing bodies, on the basis of morally relevant characteristics of citizens
(3) Civil liberties: freedom of information, opinion and expression; freedom of property, ownership and trade; protection of the private sphere and private life

To understand how and why modern liberals perceive environmental problems, we have to return to the aforementioned classical roots of liberalism. It is only there, before any such thing as the ozone layer was known to exist, that we can hope to find the unpolluted spirit of liberalism, the *Weltanschauung* that shapes (or shaped) the liberal's conception of what is, and what is happening, in the world. At the very roots of liberalism, in what I call the first phase, we find that there was no such thing as an environment *in its own right*. In Lockean times, the environment presented neither moral nor factual problems; the whole idea of an environment did not exist. Humanity did not live 'in' an environment, it lived instead on the verge of nature, just over the border. Nature had two roles to play in liberal thought: physically, it was an inexhaustible source of resources; intellectually, it was the incarnation of the laws of nature over which humankind had triumphed, which it had transcended.

A perfect illustration of both points is offered by liberal contract theories like Locke's (Locke 1973; cf. Lee 1996), Pufendorf's (Pufendorf 1991), even, to the degree that he adhered to one, Spinoza's (Spinoza 1951, 1972). In all these, humankind has a special position in relation to metaphysical nature: it differs from the rest of the universe, either enough to rise above nature (Locke, Pufendorf) or enough to think it does (Spinoza), to create a realm of its own: society. The distinguishing feature of society is that it allows humans to be (or feel: Spinoza) free, within the bounds set by the laws of nature, to shape their own life. Society itself is, in a sense, a necessary product of nature: in building a society, humans follow the first law of nature that all things follow: the urge to survive, to persist in existence (*conatus perseverandi*: Spinoza). It is therefore not as if humans had fully thrown off the bonds of nature; there are no such bonds for them. Rather, they are endowed with a special gift that allows them to discover and – all in accordance with natural law – use the leeway that nature's leashes gives them. This gift is an emotion, a 'motivational force', called reason. Reason allows them either to work out ways of bending the laws of nature to their advantage, for example by founding societies (Locke, Pufendorf), or – by means of societies – by discovering ways of transcending this very illusion of grandeur (Spinoza) and be reunited with nature.

It is fascinating to see how many of the now established elements of liberalism were already present in this embryonic phase of its development. Note, for example, the crucial role of reason: in principle, all humans have it by virtue of being human. Here lies the beginning of arguments for the political equality and influence of citizens, for the individual as the source of all political authority, for the priority of private over state interests. Note also the role of nature as a leash – the concept allows that for some the leash can be longer than for others. The more we depend for our survival on others, the less leeway we have. Here lies the source of first-phase liberalism's arguments for the exclusion of (half-wild, emotionally retarded, rationally underdeveloped) women and children, of (ditto) criminals, of employees. Here also are the criteria that liberals would later attack in their struggle *against* exclusion. Note, finally, how humankind's transcending nature by means of reason makes humans free, in a negative sense, to choose a purpose in life, a notion of the good life, a plan of life: reason has turned instincts into (as Karl Popper called it) taboo, and taboo into the absence of guiding principles. Reason alone is insufficient when it comes to sketching out the individual's good life: she has to design that herself. In this fact lies the reason for early liberalism's radical emphasis on tolerance.

What is particularly fascinating to see is how early contractarian liberals apparently perceived the part of nature outside their own realm. In every description of the 'state of nature', two aspects are balanced against one another: the nasty and brutish, and the useful and benevolent. Either one can serve to help explain the genesis of society. If humans live alone, in that unfriendly state where dog eats dog and human robs human, the urge to survive will cause a revolution and point the way towards cooperation and away from want; if they live communally in more friendly circumstances, that same urge will incite reform and show them how to better their position. Nevertheless, both aspects are present in any one contract theory; both are recognized as faces of this Janus, nature.

Nature, therefore, is two things at the same time. On the one hand, in its friendly aspect, it provides for all our wants, and it provides for them incessantly – it is an ever-flowing horn of plenty, an indepletable collection of resources. On the other, it is wild, raw, unconquered, untamed – we have to cultivate it, tame it and transform it into something edible, drinkable, wearable, smokeable, readable, in general: useful. In fact, not only *can* humankind use nature to the fullest extent necessary – it also has a 'right' to do so, a natural right in the pursuit of the first law of nature, i.e. the quest for survival. This most basic law of nature serves as the major in arguments for the legitimation of property: without the concept of 'nature', our modern understanding of property would have been impossible. It also explains the need for a social contract: namely, to protect each individual's equal (natural) right to use the resources available in the pursuit of their private conception of the best road to first survival and then the good life.

Classical liberalism recognizes only one essential distinction in nature: the line dividing reasonable and unreasonable beings. Like medieval philosophers, liberals still see a hierarchical difference between humans, animals and plants, a hierarchy often still crowned with angels and, the *non plus ultra* of reason, God. Yet this hierarchy is at best metaphysically relevant, never ethically or politically. The fact that animals stand halfway between plants and humans is not a reason to give them a special status; being unfree and beyond ethics, they are as much part of nature-as-resource as cabbages are.

The period and thought of the French Revolution lead to the birth of two new philosophies that would in time transform liberal political philosophy into what it is now. One is typically a sister of the liberal scepsis towards religious and in general 'higher' truth: utilitarianism. It rejected all notions of such moral criteria and instead (super)imposed the positivistic idea of the purely human good: the subjective experience of pleasure and pain. Even liberals who did not accept the conclusions of utilitarianism now sometimes accepted as a premise that humans and animals did have something in common, something that was morally relevant: sensual experiences. The second ideology was a typical reaction to, but in a sense compatible with, positivism and utilitarianism: romanticism, with its love for the aesthetic and emotional in general and nature in particular. It allowed liberals – particularly Americans, by the way – to make the first steps towards the recognition of the uniqueness of a landscape and its elements and so on to the idea that some forms of natural capital are simply non-substitutable. (Note, however, that romanticism itself is not necessarily either liberalism-friendly or ecologism-friendly. Its glorification of an older, more simple way of life can just as easily be turned against the modern, urban, industrial life associated with liberalism, and its particular brand of love of the natural life may support farmers and (aristocratic) hunters more than activists and hunt saboteurs.)

It is not easy to say whether liberalism absorbed elements of utilitarianism and romanticism, or whether utilitarianism became liberal and romantic. We might say that T.H. Green illustrates the first hypothesis and J.S. Mill the second. At any rate, the result was the arisal of 'positive' liberalism, social liberalism, next to a more hardliner's 'negative', political liberalism. Moreover, liberalism in both versions, though the latter less than the former, now offered room for moral consideration of the welfare of animals and the protection of nature. It is not coincidental that this period saw the emergence of the RSPCA and similar organizations (cf. Tester 1991: 88 ff.) as well as the first nature reserves. Even the unromantic consequences of progress, described at its most horrid by Malthus and experienced most deeply by those living in the polluted centres of British industry, were reflected in liberal thought. It is in Mill's *Principles of Political Economy* that we first encounter the idea of economic limits to growth and find arguments in favour of a steady state economy (Mill 1965). Note, however, that the latent tension between the endangered Nature The Beautiful and nature as an inexhaustible collection of resources remained precisely that: latent. After Mill, liberalism kept virtually silent about environmental issues.

Liberalism in this second phase is far less interested in the role of the laws of nature, either in general or particularly when property rights were at stake. Especially after the hightide of romanticism has passed, in the days of Popper and Hayek, the interest in nature wanes again. The focus of attention shifted to political and economic matters, to the threat of totalitarianism and so forth. It was typically committed to a more utilitarian assessment of the use of private property and a more‾ secular, even democratic foundation of rights.

3.3 The greening of liberalism

In the course of the 1960s, liberal political philosophy was awakened like a sleeping beauty by the kiss of Prince Charming, the legitimation crisis of that time. Until that moment, political philosophy – at least in the Anglo-Saxon world where liberalism had its den – had been dominated by the analytical linguistic approach. In 1972, one of the people who contributed to the development of a liberal response to the legitimation crisis, the until then virtually unknown John Rawls, published the book that would reshape the outline of liberal and indeed nearly all political philosophy, *A Theory of Justice* (Rawls 1972). Soon afterwards, it was already described as the authoritative statement of the agenda of political philosophy (Nozick 1974), and by 1980 Rawls had been promoted to the rank of 'saint' (as allegedly said by Thomas Nagel). A new paradigm had emerged: the Rawlsian tradition, also known as liberal egalitarianism (Kymlicka 1990).

Although any selection is arbitrary, we could say that among Rawls's critical admirers, Bruce Ackerman (Ackerman 1980, 1983), Ronald Dworkin (Dworkin 1978, 1985), the early Brian Barry (Barry 1965) and Thomas Nagel (Nagel 1986, 1991) stand out as the most renowned representatives of liberal egalitarianism. Robert Nozick (Nozick 1974), James Buchanan (Buchanan 1975) and David Gauthier (Gauthier 1985, 1986) are Rawls's best known libertarian opponents. Together, they represent the liberal *con*ception but, since they actually had little conception of it, even more so the liberal *re*ception of the idea of the environment as it emerged in the 1960s.

We notice two things when we consider this – at least in the Anglo-Saxon world – mainstream paradigm in political philosophy. One is that two of the three central topics of debate are extremely familiar: the issue of the toleration of different lifestyles and, due particularly to Nozick's critique of Rawls's solution to the former problem, the issue of (self-)ownership rights. Being liberals, even though with diverging tastes and different deeper reasons, all mainstream authors defend the view that a moral standard for the common good should not play a dominant role in the design and institutions of a society, even if there were one or if it were possible for one to be discovered.[10] It is instead the task of the political institutions of society to safeguard and sometimes promote a both maximum and equal freedom for individuals to pursue the life (in Rawls's authoritative terms: reasonable plan of life) of their choice and the good of their choice.

At this point, a third topic of debate is introduced: distributive justice. Where the idea of equal freedom is translated in positive terms as equal opportunities, equal resources, equal welfare, in general as equal *meaningful* rights – an idea inherited from nineteenth-century liberalism and socialism — the need may arise to (re)distribute the fruits of the unfair initial distribution in life of talents and handicaps. And here, almost automatically, the question of the legitimacy of (self-) ownership arises: how certain are we of our rights if the least inequity in our own or other people's innate capacities may void them? Why should the (disputable) fact that the fruits of our labour may be (individually) undeserved imply that these fruits automatically become collective property? Does not redistribution imply taxation, and would taxation not mean taxing talents rather than efforts, since the talented will either be forced to work for others or refrain from using their talents? Is not distributive justice then at the same time a kind of slavery, a kind of collective ownership or ownership by the weak and lazy of the talented and active?

Another noteworthy point with regard to mainstream liberal political philosophy is that it discusses (what were and in most cases still are) current affairs – the kinds of issues that attracted the attention of the general public rather than the academic world. The list of popular political questions included international peace, international justice, women's rights and also, for the first time, the environment. I must immediately amend this, though: at least the environmental questions that were discussed, even though related to issues that are now recognized as environmental issues, were neither induced by a genuine environmental awareness nor were they intended to address the agenda and concerns of environmentalists and ecologists. I do not think it is exaggerated to say that the 'environmental' concerns of mainstream liberals were limited to the following four issues only: pollution, animal rights, the global distribution of resources and obligations to future generations (cf. for a similar conclusion Taylor 1993). And neither of these four was seen as a typically environmental problem or even as having an environmental dimension.

Consider first the discussion of animal rights. The aim of authors like Nozick and Ackerman in discussing the status of animals had nothing to do with bio-diversity or the imminent extinction of species, nor with the distribution between human and animal of access to resources, nor with the treatment of individual animals, nor even, finally, with an attempt to prove that humankind would be uniquely fit or destined to own and use nature. Instead, animals served the intra-theoretical purpose of marking off humans as one category and identifying the traits that would make humans and human interests morally and politically rel-evant. Animals served to illustrate and support the principle of political and social equality. The remarkable fact that distinctions were less easy to draw than might be hoped did, however, result in the recognition that humans could actually have moral obligations towards animals – a discovery that would, in time, make the issue of animal rights itself fashionable in philosophical circles.

Some liberals have written if not in support of, then at least about, the popular perception of an unjust distribution of power, wealth, the standard of life, oppor-tunities and access to resources on a global scale. In recent years, this same topic has been put on the green agenda, where it has been recognized as a social, political and economic problem, every possible solution of which will have far-reaching environmental consequences. Not so in 1970s philosophical liberalism: there is no attention for or recognition of the environmental dimension. Even more dis-tressing, from the ecologist's point of view, is the fact that whatever solutions liberals defended were solutions between nations. The sovereignty of nations was no point of debate; it seemed that there could be no reasons to overrule national interests, to overcome the division of the world in states as elementary units or to independently justify supranational, even global, policy-making institutions. (Cf. in particular Rawls 1972.) Furthermore, (post-)Rawlsian liberals have dis-cussed, actually at great length, one of the great issues in green discourse and one of the strongest grounds for environmental concern and action: our obligations to future generations. Yet the environmentalist will again be disappointed when discovering in precisely which way the topic was addressed. In Rawls's *A Theory of Justice*, the existence of obligations to (some) future generations is taken for granted, and in fact few have ever dared to question this assumption, but the resulting 'savings principle of justice' was typically designed with the idea of perpetual economic growth at the back of the mind; Rawls made no provision for diminishing growth or worse. In fact, it seems that his savings principle only recognizes obligations to future generations in fair weather. Other liberals spent even fewer thoughts, though perhaps more words, on future generations (Nozick 1974; Ackerman 1980).

Incidentally, future generations are also the subject of an at first sight rather academic debate on population size inspired by Derek Parfit's work in this field

(Parfit 1984). What makes it look academic (as always: to greens) is that it bears no direct relation to the green issue of overpopulation and population control. Instead, it deals with the question whether we have a duty to produce happy offspring rather than simply offspring, whether causing people to exist or causing them not to exist can be intrinsically good or bad, and how the size of future populations should be related to average or overall happiness. All this may indeed seem rather abstract, and yet, as I shall argue in Chapter 6, Parfit's ideas will probably be more welcome to greens than the views of non-utilitarian liberals who must focus on the consequences of population policies at the level of individuals.

Finally, in the last 20 to 30 years, pollution seeped in as an interesting topic for liberals – especially, I should like to stress, in the unlikely regions of libertarianism (Rothbard 1973; Nozick 1974; Buchanan 1975). Yet once more, the approach to the problem is thoroughly un-green. What is at stake for libertarians is the possibility of a trade-off between a hypothetical right to pollute as a consequence of the unquestioned right to manufacture goods, and an equally indisputable right on the part of victims of pollution not to have their freedoms infringed by the activities of neighbours. Notice, however, that the libertarian presumption is that pollution is an illegitimate burden that *can* somehow be compensated for, and that in fact the possibility of compensation for pollution is a special case of the almost universal faith in the intersubstitutability of resources. Notice, moreover, that pollution itself does not seem to be the problem (since it can apparently be counterbalanced); the problem lies rather in harmonizing incompatible rights.

If we want to assess the contribution of (post-)Rawlsian liberal political philosophy to the debate on the environmental crisis, we must take account of the good as well as the bad and ugly. It is undoubtedly true that liberals have, up to the 1980s or even 1990s, either forgotten or neglected most topics of green political debates, and that they have taken up others only with the greatest reluctance. For one, the few environmental problems they discussed were not recognized as such. For another, the whole range of newly developed green ideas seems to have passed by unnoticed: there is no mention of the intrinsic value of nature, no notion of the possibility that economic growth may end one day, nor of ideas like the depletion of resources, physical limits to growth, sustainability or carrying capacity – not even of the protection of nature reserves. Frankly, it seems as if the expulsion of metaphysics from twentieth-century 'decent' academic philosophy had totally eliminated the metaphysical concept of nature that was so prominent in classical liberalism (cf. Stephens 1996: 4, 11), and that the radical separation between philosophy and empirical science had done the same for the physical phenomenon nature.

On the other hand, no matter how marginal the *questions* may have been with which liberal political philosophy wrestled, the *answers* it formulated do in fact have very profound and radical consequences. For example, it is now virtually impossible for a philosophical liberal to design a metric of distributive justice that does not at least address the issue of the interests of future generations, even in bad times. Moreover, liberals have opted for what is sometimes called the similarity approach in ethics, according to which the moral relevancy of a subject depends on its having relevant traits in common with a reference group (immigrants with foreigners, animals with humans, etc.).[11] Apart from serving as a means of supporting equality of humans and the equal concern due to them, this opens the door for arguments in favour of the protection of non-human nature – and these arguments do not necessarily have to be utilitarian in nature. Next to this, the debate on the Parfit People Problem may well have provided ways to discuss the extremely sensitive and emotional issue of population policy. And finally, with pollution liberalism has recognized the existence of burdens other than thieves and tax persons, thus opening itself up to questions of substitutability and, ultimately, sustainability.

The one great exception to the rule that liberal political philosophers are not intrinsically interested in environmental matters is Brian Barry (Barry 1989, 1993, 1995a, 1995b) whose work for the last decade has focused both on the – among liberals – fashionable green issues, albeit at greater length, and on some of the less popular questions: the limits to growth, the anthropocentric versus ecocentric approach to the value of nature, and the related question of whether nature has intrinsic or instrumental value. Recently, Barry has started on a new project: the design of a liberal theory of distributive justice for a sustainable society. For a long time, Barry seemed to be the only established mainstreamer who took ecological issues seriously. In the last three or four years, however, he has at long last been joined by other established figures like John Gray and David Miller. Moreover, John Rawls's reinterpretation of his own theory of justice in *Political Liberalism* includes a new justification of his savings principle for justice between generations (Rawls 1993: 274) in terms of mutual advantage rather than the (contingent) love of people for their own offspring. As I show in a later chapter, Rawlsian liberals may well discover that the implications of this on the surface minor change in Rawls's view on justice between generations fundamentally changes the focus of mainstream theory in ways that make it far more open to environmental issues. Now that green liberalism is no longer the interest of just one eccentric, Phase Four has begun. And yet, some things are still missing.

3.4 The green problems of liberalism

Liberalism has come a long way in accepting the need to address environmental issues, but it has taken quite some time to do so, and it still has a long way to go. If we want to assess the progress made so far, we must first recognize that the debate on the implications of the limits of growth thesis and its policy principle, sustainability, has only just begun. In this section, I consider the directions in which the debate on the greening of liberalism is likely to develop.

First, there is the issue of sustainability. During the past decade, debates about environmental issues have been increasingly put in terms of sustainability, so much so that nearly everyone on the spectrum from very shallow green conservationism to very deep Green ecologism now prefers to use this term rather than anything as politically sensitive as ecological, environmental or green. It must be admitted that the notions of sustainability and its dynamic counterpart, sustainable development, are very well suited to proliferation on such a wide scale. The often quoted and more or less authoritative definition of sustainable development in the Brundtland Report is open enough to allow for diverging interpretations:

> to ensure that it meets the needs of the present without compromising the ability of future generations to meet their needs. (World Commission on Environment and Development, *Our Common Future* (1987), quoted in Nelissen, Van der Straaten & Klinkers 1997: 282)

Moreover, those diverging interpretations are themselves in turn open to further interpretation (cf. e.g. Eden 1996: 158 ff.; see also Dobson (forthcoming) for a very lucid in-depth analysis of 'sustainabilities'). We should, however, not lose sight of the fact that sustainability is not a non-commitant term; it still has some substantive meaning. This substance, the idea that there are limits to growth and obligations to future generations, is exactly what is needed to put environmental issues on the liberal agenda: without these physical limits, only moral reasons could draw attention to the relation between humans and environment – and liberal neutrality might well consider the latter an issue for private rather than public deliberation. Then again, even this core idea can be interpreted in several ways. One relates to the notion of the Earth's limited carrying capacity: there are only so many people it can feed, there is only so much pollution it can absorb. Another is in terms of economic development: there is a moment at which the costs, including or even excluding external environmental costs, of producing goods will rise above what a national economy can afford. Yet another points to the social limits to growth: there is only so much progress, change, flexibility, artificiality and alienation a society can bear. And then there is the geological interpretation: there is a physical limit to the amount of resources that can provide the energy for production, a limit to the amount of resources that can be transformed into goods, and a limit to the achievements that can be made in the field of increasing the efficiency of production processes. These and other interpretations have still not been adequately differentiated, nor have they been introduced separately in liberal discourse.

We may also expect the introduction of the notion of limits to growth and resources, and with it that of sustainability, to lead to questions of a substantive normative nature. A sustainable society need not be one big Yellowstone Park – we can imagine a worldwide version of Holland stuffed with cows, grain and greenhouses, or even a global Manhattan without the Park to be as sustainable and for many among us as pleasant as the first. Hence, a greener liberalism will have to define more clearly what kind of sustainability, what kind of world, it aims for.

Next, modern liberalism is still anxious to avoid any attack on the sanctity of individual plans of life, but this issue may well become unavoidable. Usually, the green view of sustainability recognizes three aspects: a supply side (resources), a demand side (consumers) and the distributive structure of society (see Section 1.6). To warrant sustainability, all these three aspects are relevant. The second aspect suggests solutions in terms of population size, but it is notoriously difficult to justify binding measures from a liberal point of view, as we shall see. The third suggests that we look for fair distribution schemes that maximally satisfy legitimate demands. Any solution, any set of distributive principles Q and criteria for the legitimacy of demands P that we were to propose, should be general in two senses: it is to be applicable to any morally permissible permutation of the set of possible conceptions of 'resources' (parameter 1) and to any corresponding permutation of 'consumers' (parameter 2). But the problem is we can never guarantee that there will not be too many people to satisfy even their most basic needs, let alone to satisfy these fairly. Hence, liberals seem destined to depend on one of two remaining strategies. The first is that of technological and geological solutions to the resources problem (parameter 1), a strategy that is both undependable (unpredictable) and questionable, at least by the standards of many environmentalists, who claim that uncritical reliance on technocratic solutions is what brought about our present environmental problems in the first place. The one remaining alternative is to preach austerity, i.e. to restrict the number or amount of people's preferences, desires, elements of plans of life, or whatever we want to call them. Since one of the main aims of liberal democracy is to promote the liberty of life rather than restrict it, this may also prove to be a hazardous route to salvation. At least theoretically, green liberals may therefore be thrown back after all on either population control or on ways of influencing the genesis of preferences, the way in which people's desires and plans of life are generated.

In the context of this problem, we should also consider the unique nature of environmental problems. One of the odd things about sustainability and the more general concept of environment is not just that it can serve to bring the most diverse problems under the same heading (see the next section) but that it actually *is* made up of all sorts of interconnected problems. At the very least, environmental problems consist of waste material passing on from consumer to soil to water to sea to air and on to natural resources or straight on to the consumer. Problems of sustainability cross all borders between media, ecosystems and political entities. The interconnectedness of environmental problems thus further limits the possibility for liberal democratic regimes to warrant sustainability – and it raises the question whether the division of responsibilities over states is compatible with policies aimed at sustainability.

Then there is an issue that so far has escaped the attention of both greens and (most) liberals, due to, on the one side, the green preoccupation with what has to be done to ensure the survival of humankind and the planet, and on the other side the trenches that have been dug between liberal egalitarians and libertarians. The issue I am referring to here is that of the legitimacy of property rights. Environmental issues make us aware, more than ever, of the shaky foundations on which the legitimacy of property (whether collective or private) rests. Consider the so-called Lockean proviso on which, in the grey past, the now unquestioned legitimacy of property was founded: a property right to a resource x requires that 'enough and as good' of x is left for others. The Lockean proviso (Locke 1973) presumes that resources are by definition abundantly available, whereas modern variants of it, as defended by e.g. Robert Nozick (Nozick 1974), presume that there can be abundant (in a sense infinite) compensation for stock reductions, which in turn presupposes the possibility of infinite recycling. The limits to growth have brought these two (and other) theories of legitimate acquisition into really quite serious difficulties. If resources are no longer abundantly available and if the modern alternative, abundant compensation for stock reductions (Nozick 1974) is also excluded since it presupposes the possibility of infinite recycling – then what remains?

The point to note here is not just that these problems may only apply to other (almost forgotten) justifications of private property – it is also that they apply to justifications for property in *general*. If we can share Robert Nozick's observation that the absence of a justification for private ownership does not automatically justify collective ownership (Nozick 1974), we see that any attempt to defend collective ownership will meet precisely the same problems. (For an evaluation of Locke's proviso as an alternative route towards greening liberalism, see Dobson (forthcoming).)

Some liberals have also affirmed the existence of a problem in regard to the source of value (i.e. intrinsic/instrumental) and in regard to its attribution (human, mammal, animal, plant, rock, landscape). Yet few have seriously discussed it so far. This is odd in view of the immense effects that the recognition of even a very limited degree of moral relevancy of mammals must have on political (e.g. distribution) decisions – not to mention the repercussions of the recognition of the intrinsic value of all nature, as deep ecocentrists would have it.

Finally, there is an urgent need for a metadebate on green liberalism. The four relatively neglected topics mentioned in this section, plus the recent contributions mentioned in Section 3.3, do not seem to form a coherent story. So far, it is more like a set of possibly mutually exclusive amendments to the existing body of literature on liberal distributive justice. What is needed now is a decision on which topics should receive the greater part of our attention and which should be set aside for a while as being of secondary importance. In making this decision, we should for once not be guided by what is theoretically interesting nor by the mood of the day, by what the competitor, ecologism, propagates as urgent. The green agenda for liberals has so far been dictated by environmentalists and hence determined by an ecological, not a liberal, ontology or *Weltanschauung*. What green liberalism needs most of all is a perspective of its own, or at the very least a critical reassessment of the ecological ontology. We must also decide which solution strategy to follow: the inductive route of dealing with separate questions until some kind of *status quo* has been established, or the deductive route of first designing grand theories of sustainable or ecological liberalism and only then applying them to particular questions. The first route seems more practical (it is in fact the route I follow in this book), since any answer to any of the relevant 'particular' questions can immediately help philosophical liberalism confront eco-logical challenges. Nevertheless, it would be a great help if a second Rawls appeared with an agenda-setting *A Theory of Green Justice*.

3.5 AIDS, women and deforestation

The title of this section is first a tribute to two female colleagues who one day had to listen to a male colleague describing the three major problems of our times as 'AIDS, women and deforestation' and had trouble convincing the speaker that his was not the luckiest choice of words. It is also, secondly, a good *pars pro toto* label for the subjects on the green (activists') agenda: in principle, virtually every-thing is on it, since virtually every issue can be given an ecological twist. Smoking, for instance, is an environmental issue because it involves the creation of biological resources, their transformation into consumer goods and subsequently into waste products. Sport is an environmental issue because it involves the use of space for facilities that would perhaps not be necessary if we lived a more healthy life. This book itself is an environmental problem unless the efforts that went into writing, printing and publishing it can be proven to be, on the whole, beneficial to the environment – provided we first establish that environmental consequences can be measured in terms of costs and benefits. Racism is an environmental issue, since it involves unequal access to resources and discrimination of possibly more environmentally sane cultures and habits. Even pure reflection can be an environ-mental issue – it is reflection that brought humankind from the stone age into the age of computer technology. Not every subject *has* to be on the green agenda: not every greenish confession requires faith in the all-permeating omnipresence of the Green Idea. Nevertheless, everything *could* be on it.

There is a kind of shortlist of environmental problems on which all environ-mentalists and ecologists seem to agree. These are what I call first order problems. They relate to purely physical problems in which the hand of humankind is visible as a cause and effect become visible as possibly irreversible changes in nature. Disagreements begin when second order problems are introduced, i.e. when the search for explanations and deeper causes begins. It is the exact mix of explana-tions in terms of social and political circumstances, moral ideas, technological modes of production, discourses and subtexts, epoche and paradigms that deter-mine, first, which other topics could be added to the list of first order problems, and second how first order problems are thought to arise. On the list of second order problems, we find items like uncontrolled economic growth, inefficient use of resources, Third World poverty, military expenses, population growth, increased use of energy and increased consumption in general (luxury goods as well as food), and deeper down moral deficiencies like greed, lack of concern, anthropocentrism, materialism, consumerism, lack of spirituality, a temporal bias in favour of present generations.

We should also distinguish a level of third order problems to explain second order and hence first order judgements – which implies that even the shortlist of first order problems is contestable. First and second order problems cannot be recognized, let alone ordered, without a deeper theory, a *Weltanschauung* that makes them problems. The shortlist of first order problems which I am about to introduce, for instance, lists issues about which an orthodox stoic probably could not care less. But then, greens are not stoics. Here, I shall ignore second and third order differences of opinion. The shortlist, which is derived partly from Albert Weale's inventory (Weale 1992: 4) and partly from Bosselmann's (Bosselmann 1992: 344), is intended to reflect a *communis opinio* among greens, not a general truth. I mention them not because they are problems but because they are perceived as problems.

Resource management and subcategories
- depletion of non-renewable natural resources (energy and others)
- extinction of animal and plant species
- eutrophication
- deforestation
- erosion
- desertification
- water shortage
- damages to ecosystems
- diminishing biodiversity
- treatment of animals

Pollution and subcategories
- poisoning of flora and fauna
- soil
- groundwater
- rivers
- seas and oceans
- air
- acidification
- ozone depletion
- global climate change (e.g. greenhouse effect), subdivided into:
  - global warming
  - rising sea level
  - changes in flora, fauna and ecosystems as a whole.

Not all these problems are necessarily relevant from a liberal point of view, at least not initially. What constitutes a problem for liberal democracies is determined by the list of criteria developed in Chapter 1. On that basis, resource management and pollution issues will only be real problems if they currently endanger the process of reconciling and satisfying individual preferences. It is not until we can establish that liberal democratic systems have to provide opportunities for their continued existence that the time horizon can be expanded to include the future and from there on perhaps future generations and non-human interests.

To us, reflecting on the possibility of green liberalism, the topics on this list are all relevant, albeit in a different way. They can serve as test cases in assessing how green green liberalism can be in the eyes of greens. Yet at the same time, it is far less important to find a liberal democratic answer to these concrete environmental problems than it is to determine the positions liberalism should take on the more abstract questions introduced in Chapter 2. Answers to the latter type of question determine whether practical problems are problems at all, and in which sense they are problems − from a liberal point of view. It is only then that it makes sense to discuss liberal democratic solutions to environmental problems. After all, solutions to practical problems demand rules and criteria for success; otherwise environmental policy would be guided by chance. Philosophy is prior to good policy.

In dealing with these theoretical questions in the next six chapters, I shall let myself be guided by the notion of sustainability or sustainable development. Sustainability helps us to order our questions. As noted before, it is built on two premises: that future generations are recognized as relevant, and that the existence of limits to growth are recognized. It also organizes the choice of solution strategies into three groups: supply, demand and distribution. I shall use these ideas as working hypotheses to see if liberal democracy can be sustainable, thereby determining which positions liberals should take on the green issues of Chapter 2, and only then, having established that liberal democracy can be sustainable to a considerable degree and that it *must* be aimed at sustainability, discuss whether this will also make liberalism really green.

It follows that it is not my primary objective to find solutions to practical environmental problems. My aim is to indicate that solutions are possible. Consequently, I shall only discuss some of the issues on the shortlist, I shall discuss them only incidentally and only to the degree that they pose problems of a politico-philosophical nature.

3.6 Intermezzo

Having reached the end of three chapters of often exhaustingly abstract groundwork, this is a perfect moment to sit back and rekindle the spirit. In this section, I want to recapitulate what I set out to do, what (little) has been done so far, and what still awaits us.

I set out to discover whether liberalism could go or be 'green'. The foundation has now been laid and the results can be summarized in a few paragraphs. First, I have given a description of the normative, prescriptive or moral aspect of liberal democracy. It was not a description of how liberal democracies actually work, nor one in terms of the ideals that *should* lie behind practice. The former would tell us little about what the morality of liberal democracy *could* allow, the latter depends on the special perspective that alternative schools in liberal political philosophy, from libertarian to social democratic, hold to be right. My model was instead formulated in terms on which these schools should be able to agree: that is, in terms of the ideals that *actually* lie behind liberal democracy and at the same time *actually* describe the moral minimum on which liberal theorists agree in principle. This part of our analysis resulted in a list of 'stringent conditions of liberal democracy'.

Next, I mapped the field of green, environmental or ecological theory. I distinguished four levels of green debate (metaphysics, ethics, politics and policies), described the issues and possible positions at each of these levels, and deselected some positions that turned out to be prima facie incompatible with the conditions of liberal democracy. Hence, we have delineated an area, or in green terms discovered an ecological niche, within which 'green liberalism' might find the right soil to flourish.

Last, I considered the evolution of liberalism itself, to find that it lost most of its metaphysical interest in questions of nature over the course of three centuries; that it gained some new ideas with regard to more profane questions of an ethical, political and policy nature; and that it nevertheless remained silent on green issues – until very recently. Given that there are good reasons to develop a green perspective for liberalism, given that there is (virtually) none so far, and given that there is room for such a view, I formulated a green agenda for liberalism. I made two lists of issues with which green liberalism will have to deal. One concerned green issues that might ask for fundamental changes in liberal theory, the other specified the purely physical environmental problems to which a viable green liberal political theory and system should have some kind of answer.

In the following six chapters I shall try to deal with this green agenda, following the four levels of debate described in Chapter 2, but postponing most metaphysical issues until the very end. Chapter 4 discusses ethics. In particular, it deals with abstract questions of value: what is value, what are its origins, is it or can it be intrinsic to an object or entity, external or both? Liberal democracy's attitude towards nature will vary depending on the answers to these questions, from brute exploitation if the value of nature depends on human preferences to scrupulously respectful trusteeship if it has intrinsic value. Or so it would seem. I shall however argue, among other things, that the notion of intrinsic value should be abandoned, possibly for theoretical reasons but certainly and at the very least for practical reasons. Maybe the idea that nature has intrinsic value can be reduced to a form of external value, but it can certainly be represented as such, as something that is valued by those who either create or discover value in the world: individuals guided by their plans of life.

Chapter 4 also deals with two issues that immediately follow from the discussion of the nature of value. One is substitutability, i.e. whether or not one 'vehicle of value', say an old chestnut tree, can be replaced by others, say, a painting or hologram of the tree in a museum, a load of furniture, a series of poplars. I argue that a conception of the value of (elements of) nature as having external, even instrumental value cannot in any way serve to discard the notion of substitutability, but neither can it exclude the possibility that some or even all parts of nature are insubstitutable. The other question is 'who counts?' – which entities deserve which kind of protection against which kinds of interference with their existence? The latter question will be answered in terms of formal rights (cf. Chapter 1). Roughly and slightly romantically put, my conclusion will be that as long as there are creatures for whom parts of nature are experienced as insubstitutable, liberal democracy has a principal obligation to protect nature against any kind of destructive exploitation.

Chapter 5 is about politics and discusses principles of justice or, to put it very formally, the principles that should rule the acquisition, attribution and distribution of (formal) rights by and over the recipients governed by liberal democratic political institutions. The two most important ideas I introduce here are (1) trilateral obligations between previous, present and future generations and (2) the restraint principle. The first is a special way of representing the notion of obligations to future generations; in fact, it is a modification of one of John Rawls's recent amendments to his original *Theory of Justice*. It connects the survival of society and social cooperation to the idea that society is a scheme of cooperation among different generations existing next to one another, and combines this with an obligation to treat the next generation no worse than you would want the previous one to have treated you. From this and a few auxiliary theses, the restraint principle for the distribution of formal rights follows, according to which every physical object should be dealt with as if it were insubstitutable, at least as much as humanly possible. Against our expectations, this leaves us – still within a liberal democratic frame of thought – with a far better protection of nature or natural capital against exploitation, substitution and extinction than could have been expected.

Unless the usual 'other things being equal' condition obtains, all this can never guarantee sustainability, let alone the viability of greener versions of society. In Chapters 6 to 8 I drop this condition and discuss policies aimed at controlling the three parameters of sustainability: demand (Chapter 6), distribution (Chapter 7) and supply (Chapter 8). As to questions of demand, I mainly focus on population control as the one and only way to safeguard sustainability. I present the rapid reproducer paradox, according to which population policies always punish the good guys and reward the guilty parties. After ascertaining that morally justifiable demand-side policies can never guarantee sustainability, I turn to just distribution schemes as alternative solutions (Chapter 7). Here, I take up the restraint principle once more to consider its effects for just acquisition, possession, distribution, transfer and use of nature. I discuss the consequences of these principles both within and between political institutions – thus, in a way, addressing the green concerns about international justice. Finally, Chapter 8 deals briefly with supply-side solutions (cf. most issues on the green agenda in Section 3.5). We shall encounter problems here quite similar to those in the field of demand-side solutions. As a result, we find that liberal democracy must paradoxically combine adherence to the restraint principle with active support for technological solutions to environmental problems. In the end, then, there is bad news and good news. The bad news is that liberal democracy cannot guarantee sustainability for the full 100 per cent. The good news is that, in environmental terms, liberal democracy is able (morally) to do the best that can be expected from any political system.

Chapter 9 consists of dessert and leftovers. It opens with an analysis of the metaphysical positions liberal democratic theorists could, or would have to, accept if they stayed within the boundaries of sustainability. As it turns out, there are only a few options liberalism excludes: it is inconsistent with mysticism, materialism, determinism (to some degree) and an understanding of nature as being naturally in equilibrium. Beyond this, anything goes: liberalism could even have a deep green, holistic and organicistic view on nature. However, I shall argue that as a rule, liberalism's metaphysical stance is of no consequence whatsoever for the green political potential of liberal democracy.

A second issue I discuss here is whether it is possible and desirable for liberalism to move 'beyond' sustainability: that is, towards more typically deep green conceptualizations of environmental problems, towards a less conservative view, or both. The first would, among other things, require that obligations to reference groups other than future generations of humans be recognized more explicitly than has been done so far, and I argue (once more) that this is impossible. It would also require that a more eco- or biocentric, more distinctively holistic, less materialistic, less egoistical, less technocratic ethic be incorporated in the institutions of liberal democracy. I shall argue that even if this were possible, it would most likely be ineffective, and if it were effective, it would still be undesirable. Liberal democracy would then cross the line between democracy and authoritarianism. The second modification, the move away from conservatism, requires that we drop some or all references to limits to growth. However, calls for a less conservative attitude towards nature and towards social development than sustainability prescribes are partly based on a category mistake. Both the precepts of liberal democracy and those of sustainability demand that purely *physical* limits to the development of society and nature are morally relevant and should be taken seriously. Yet there is nothing in either sustainability or liberal democracy that puts the then (perhaps) imposed *moral* limits, i.e. limits on what we should do rather than can do, beyond critique.

I then discuss the genesis of preferences. Up till now, I have assumed that preferences are given and that the political process is an input–output machine. Environmentalists and theorists of democracy, however, point out, with good reasons, that democratic processes and debate shape and transform preferences and opinions rather than merely straightforwardly translate them into a collective choice. In my comments, I make two points here. I shall assert that there is no guarantee and no reason to believe that more democracy or public deliberation will make a society more environmentally sane, as some environmentalists have argued. I defend the view that environmental interests are still best protected when they are, to some degree, put beyond debate and protected by principles like the ones developed in this text. I shall also try to give a short answer to the objection that liberal democracy, being procedural, is incompatible with the ends-oriented approach of environmentalism (cf. Dobson 1996). I hope to have shown in the preceding eight chapters that it is not.

Last, I turn to the most fundamental green critique of liberal democracy, the one that sees it as a vehicle of materialism, consumerism, egoism, modernism and modernity, industrialism and many other deeper causes of the ecological crisis. Although critiques of this sort are well founded in so far as liberal democracy is in many ways an *historical* product and facilitator of said deeper causes, we must make an *analytical* distinction between liberalism and the Spirit (or Spectre) of Progress. Liberty, equality, justice, democracy and all other values incorporated in the institutions of liberal democracy are not only the monopoly of one cultural 'epoche' (in Michel Foucault's sense of the word). We must also distinguish political from economic liberalism, since it is in the latter sphere that materialistic, consumeristic attitudes and behaviour find their most extensive freedom from public and moral scrutiny. We should even distinguish between the idea of a free market and its present incarnation in the form of capitalism. As an example of an alternative form of the free market, I shall briefly discuss the way it was conceived of in medieval times. We should not read this as a plea for a return to medieval society or as one for communitarianism. All the example demonstrates is that green liberalism requires individual responsibility to be viable: citizen virtue but even more civic virtue, and particularly virtue in the market place.

4

On value

4.1 Intrinsic and external value

We crave, desire, reject, favour, disapprove of things. We do not like the colour of this, the taste of that, the behaviour of one and the attitude of another. We pass value judgements, value as a noun being for normative standards (norms) what 'unit of measurement' is for empirical standards like size, weight, wave length, left–right, income, education. We judge by standards of taste, beauty, rightness, goodness, grammatical correctness. Some of these standards can be applied universally or almost universally (e.g. beauty and goodness), though they need not make much sense in all contexts. We can ask whether it is *good* that lemons are yellow or whether they would not be more *beautiful* if they were purple, but (genetic engineering aside) there is little we can do about it except perhaps complain about the laws of nature or about their incompetent implementation. This is what – to most of us – makes questions of this sort different from the

91

question whether we like the taste of lemon or appreciate the beauty of Goya's war sketches. We cannot walk away from the fact that lemons are yellow or that Goya's sketches are horrifying, but we can walk away from the sketches and stop eating lemons. Yet the gap between what we would wish to do and what we are able to do does not necessarily make the issue of purple lemons totally irrelevant. It may be possible to bridge the gap. Purple lemons are only contingently impossible, just as peace in our times is, or a generous bank, a competitive tax administration, a user-friendly cigarette. Purple lemons may be a thing to strive for. And even the truly impossible can be a subject of sensible judgements: the Third Reich should never have existed, Eve should never have listened to the snake, Don Quixote should have aimed higher.

What makes the application of specific normative standards relevant depends not on the possibility of change but on the possibility that change may be desirable. Then again, some types of change may be more desirable than others, depending on social context and convention as well as personal conviction. We value (rate) lemons higher or lower than pork according to our standard of taste (or price, or kosherness) when we decide what to eat, by colour and shape (or symbolic meaning) when we decide what to paint, by price (or taste) when we decide what to buy. Which of these standards we judge most appropriate and apply – why we eat on the basis of taste rather than colour – thus depends on a metastandard, a standard of standards by which we rate (value) distinct standards. It is this metastandard that organizes our individual theories of the good, plans of life and preferences.

Value, then, is external to the object and depends on the person judging: it is in the eye of the beholder. There is, or allegedly is, one important exception to this axiom: intrinsic value. Intrinsic value – to give a first and very rough definition – is the value that is inherent to an object, act or situation regardless of whether it can benefit or harm an individual, regardless of whether an individual perceives value, regardless even of the presence of judging individuals. The existence of intrinsic value solves the most fundamental problem of greens with one stroke: if nature has intrinsic value, then humans ought to respect nature regardless of their subjective opinions, and polities ought, in principle, to protect nature against even unanimous disbelief. If nature has intrinsic value, exploiting it is by definition evil – perhaps under circumstances a necessary evil, but evil nonetheless (cf. Larrère 1997: 20).

One thing to note about intrinsic value is that it is extremely popular in green circles. It is almost impossible to find texts in green political theory or environmental ethics in which it does not play a prominent, most often supporting, role. There are, however, three other important things we should observe about intrinsic value. First, it is not one unambiguous concept of value – there are at least four versions of it. As we shall see, it seems more appropriate to speak of *independent* value, and refer to one of these four conceptions as intrinsic value in a strict sense. Second, probably all of them are redundant; it seems that they can all be reduced to a form of external value. Third, even if they were not redundant they would still be impractical. I treat these three points more or less in this order (the last two together) in the present section. In the ensuing sections, I discuss the ins and outs of external value and show it to be consistent with liberal democracy (Section 4.2). I then address green concerns about the fact that an understanding of nature as having external value only may turn nature into natural capital and make natural capital substitutable for artificial capital (Section 4.3). Finally, in Section 4.4, I examine whose judgements on values and specifically the value of nature are relevant from a liberal democratic point of view – and to what degree.

To begin at the beginning: 'intrinsic value' is an enigmatic term. It is used in different senses and contexts by authors with different philosophical points of view, obviously with different effects. Elizabeth Harlow (1992) lists four distinct conceptions of the term as used in a loose, relatively colloquial sense:

(H1) value, independent of human assessment
(H2) value, independent of the human good
(H3) value, irreducible to states of consciousness; and
(H4) value, irreducible to human interests and preferences.

A second taxonomy, described by John O'Neill (1992), distinguishes three more ontological conceptions of intrinsic value:

(O1) intrinsic value as an end in itself, as opposed to instrumental value as a means to an end
(O2) value in virtue of an object's 'intrinsic' properties (cf. Elliot 1992); and
(O3) objective value, value independent of the valuations of valuers.

Note that O'Neill improves on Harlow by trading in humans for 'valuers' (O3). If we were to attribute animals with the ability to value, then something that has instrumental value to them would, by Harlow's count, have intrinsic value.

Finally, Ian Carter (1995a), who discusses intrinsic value in the context of a debate on the intrinsic value of liberty, argues that the term is often used in too loose a sense as value independent of any other considerations. It would be better to understand intrinsic value in a strict sense as one of four conceptions of *independent* value, so as not to confuse it with conceptions of value for which the term 'intrinsic' seems ill-fit:

(C1) unconditional value: having more value than any other thing, regardless of the value of its consequences. (C1) is a special case of (C2)
(C2) intrinsic value: value in addition to the (dis)value of the consequences of a thing. Intrinsic value, and therefore also unconditional value, is opposed to instrumental value
(C3) nonspecific instrumental value: the value of a thing x as making other things possible, independent of the value of the things it makes possible
(C4) constitutive value: the value of a thing x as an essential component of an intrinsically valuable thing y.

It has been argued that intrinsic value is a fantastical, mystical, incoherent and absurd notion. By now, the reader may feel inclined to agree or at least give up any attempt at making sense of it. I would suggest that this is not necessary. Let me try to create some order in this chaos. First, observe that not all these conceptions of intrinsic or independent value are necessarily distinct from instrumental value, i.e. from its consequences or from what we can do with a thing. This is most clearly the case with C3 and C4, but it also applies to other conceptions. O'Neill's O2, for instance, seems to be the mirror image of C4: if we ascribe O2-type intrinsic value to trees because of their magnificent shape, trees in a way become a means to the existence of a valuable end, 'shape'. We may, furthermore, not like to pay taxes but taxes may still serve as a valuable means to an end, say, social justice and be valuable only for that reason (compare H4, H3). An orthodox Marxist may feel that social justice is a trick to cover up some of the injustices of capitalism, thus postponing the inevitable realization of the human good – yet they may still find taxes an intrinsically good means to a base end since it encroaches on private property rights (compare H2). And finally, if there are valuers other than humans, a thing can be instrumentally valuable independent of human assessment (H1).

Second, it is evident that not all these conceptions of intrinsic value can exist without valuers (human or other): H1, H2, H3, H4, O1 and C3 all make implicit or explicit reference to a valuing subject. On the other hand, O2 and C4 refer implicitly to ends but like O3, C1 and C2 they seem to allow the existence of an intrinsically valuable x without a subject from whose perspective x has value.

Third, if an object has 'intrinsic' value in any sense that allows its value to be interpreted as instrumental value, or if it has value because of the judgement of a valuer, the origins of its value are external to the object itself. If its value exists independent of valuers, its value may still originate in a source external to it. Since C1 is a special case of C2, this leaves only C2 as a description of intrinsic value that can in no way be reduced to some source of value external to the object.

Having come to this point, we are now able to reorganize our conceptions of intrinsic value and distinguish between four major necessary conditions for the existence of distinct types of independent value:

(W1) value that exists independent of its being perceived
(W2) value that exists independent of the object's instrumental value
(W3) value that exists independent of external causes; and, combining all these
(W4) value that is internal to an object and exists independent of its being perceived.

I shall refer to W4 from now on as intrinsic value in a strict sense – and note that my strict sense of intrinsic value is far stricter than Carter's. The 11 conceptions discussed above are all covered by these four new ones. Just for the record: H1 is a special case of W1, H2 of W3, H3 and H4 are special cases of W2; O1 equals W2, O2 equals W3, O3 equals W1; C1 and C2 are special cases of W2, as are C3 and C4 of W3.

Having created some order I now move on to the question of redundancy. To make sense, conceptions of intrinsic or independent value must meet two criteria. First, intrinsic or independent value must exist. We must have reason to believe that there is such a thing as independent value, otherwise it can be no more relevant to the green debate than the – from our perspective – purely theoretical distinction between summation and substraction. Second, if intrinsic/independent value exists, it must enable us to formulate moral propositions that we would find impossible to formulate otherwise. If this were not the case, independent value would merely offer a different but completely inconsequential perspective – a perfect object for Ockham's razor.

Can an object have (dis)value without being perceived as such? At first sight, this seems like the kind of question a sceptic would pose to Berkeley: do things exist if we do not see them? If the sceptic were an empiricist, they could answer their own question by kicking the good bishop while his eyes were closed, proving that they were as much for real as the kick and the kick as real as the bishop. But jokes aside, the comparison between value and existence is probably invalid. It is not a necessary condition for the existence of a thing that it be perceived or that it can be perceived in some way – but it is for having value. Value is a normative concept; it requires a norm, a standard that has to be created, constructed or somehow discovered before it can be applied, and it has to be applied to have a meaning. Creation, application and meaning all presuppose a subject: a creator, applicator and interpreter, in short, perception. Thus, without perception there can be no value. However, if I am wrong on this, nothing is lost: without at least perception (not to mention the possibilities of acting and change), value is totally inconsequential (cf. on both claims Parker 1996). Let me illustrate this with an example. There is a famous story, the Last Person Argument, which is used in countless versions to show that we share the intuition that planet Earth has intrinsic value (see Lee 1993 for one version and further references). It goes something like this: imagine that you are the last human being on an otherwise flourishing Earth, you are about to die, and you have a choice between going out with a bang and going out like a candle in the night. If you push the button in your hand, the planet will explode the moment you die; if you do not, it will go on existing as pure, unspoilt nature for ever. The usual assumption is that no one would push the button even though there will be no one left to admire the beauty of the planet or do something useful with it, and the usual conclusion is that nature has intrinsic value. Obviously, many auxiliary assumptions will be needed to rid the example of contaminating elements: pushing and not pushing the button must be a choice between two acts rather than between an act and inaction; there must be no other valuers around, say, animals, whose perspective we might choose; the last human being must not believe in an afterlife nor in a virtue-ethic that prohibits destructive behaviour. Yet in the end, the Last Person Argument will never tell us anything about the last person's values *in extremis* but only about the values of those to whom the question is put – people who are not last persons and not about to fade away. We are the ones who perceive, who can compare the two possible worlds that the last person can create, we are the ones without whom the whole story could not be told or heard or even exist. And if even we could not perceive these two possible worlds, how could we ever make a choice?

Having excluded W1 and thereby W4, we move to W2: is there anything beyond instrumental value? It is hard to see that this could be possible unless a very narrow definition of 'instrumental' were used, as a means to a definite end. There is nothing we can do with the stars at this moment except glance at them, admire them, pose deep theological questions and perhaps create a romantic background for a seduction scene. Only the last qualifies as instrumental use in the narrow sense, yet we expect to derive – to use an ugly word – utility from each of these acts. We do not know to what ends the acts of staring at the sky or meditating about infinity may lead, but we seem to gain something from these acts themselves. The stars above have a use, they are instruments in our attempts to fight boredom or depression or to relax. Anything we choose to do we do by definition for a reason, good or bad, and the reason makes the act (looking up) and its prerequisites (the stars) means to whatever end our reasons are.[12]

This apparently leaves two possibilities. One would be to assume that all value is instrumental, a means to an end, so that we may expect to end up in a *regressus ad infinitum*: everything will have value as a means to an end that has value as a means to an end, etc. (O'Neill 1992: 119). Hence, a Kantian would say, the alternative is to let the chain end with an intrinsically (W2) valuable end.

Now assume for a moment that there is no such thing as intrinsic value W2. In that case, the Kantian argument would only show that both independent value W2 and instrumental value lead to equally absurd conclusions – it will not save W2. But as an objection it is probably also wrong. One possible reason is that the chain could end with a goal that is neither instrumentally nor independently valuable because it is not a value or valuable thing at all, just something that we have to do (live) or be (alive) and can do or be in different ways, thus giving rise to choices and room to moral codes. I shall not pursue this argument here; what matters is that we have no *a priori* reason to believe in the necessity of an infinite regression of instrumental values or in the impossibility of an ethics based purely on external value.

On the other hand, if we assume that such a Kantian end in itself, an ultimate independent value at the end of the chain of instrumental values, exists, we end up with an unworkable solution. Either the ultimate value is something intrinsic to the valuer (Kantian autonomy, dignity, freedom, etc.) or it is external. It is only in the latter case that nature can have independent value – but it has been pointed out that, on this interpretation of intrinsic value, all parts of nature necessarily have intrinsic value. Consequently, any attempt to choose between, say, saving an antelope and starving a lion will end in deadlock: this type of independent value does not allow for ranking, for judgements of 'more' or 'less' value (cf. Rosenthal 1996; Regan 1992; see also Sterba 1996).

Finally, can there be value independent of external causes (W3)? Can anything be good in itself, not good *as an end* in itself but straightforwardly good in itself?[13] Again, it is hard to see how. Our analysis so far gives us reason to believe that whatever value an object has must be 'put into it' by valuers and will always be open to interpretation as instrumental value. It is only the first element in a finite chain of instrumental values that could (but as we saw, need not) have value independent of external causes. Given the need for a valuer, this first value will quite likely be a 'subjective' value, one that binds all values to the valuer's sense of personal worth. Be this as it may, if we assume the existence of a thing x that is good in itself, one of two propositions is true. Either it is possible to represent x by imagining a perceiver who would value x since it is intrinsically valuable, or if we cannot represent x in that way, no one would be obliged to respect its value. In the first case the W3 type of independent value can be reduced to W1 and becomes redundant, in the second it is totally inconsequential for actors in the real world.

In summary, we have seen that three necessary conditions for the existence of various conceptions of independent value are contradictory, inconsequential or redundant – and all these objections apply to intrinsic value in a strict sense (W4). There can be no value without a valuer, and there is no reason to believe that the category of independent value cannot be reduced to instrumental or at least external value. Even if an object or act x could have intrinsic value, this would be inconsequential without its being valued by those who either create or discover value in the world: individuals guided by plans of life or theories of the good. To have more than purely theoretical consequences, intrinsic value must be recognized as anthropogenic, i.e. a valuer must 'put value into' an object. Thus, intrinsic value is, for all practical purposes, redundant and reducible to external value.

4.2 In the eyes of the beholder

In the previous section, I argued against attributing intrinsic value to nature on purely analytical grounds. First, the concept of value does not make sense without the existence of valuers: value is only in the eyes of the beholders. Hence, value cannot be intrinsic in the sense that it exists independently of the act, situation or object that has value. Furthermore, value is external to whatever we consider valuable. Even properties inherent to an object cannot have value without being recognized as valuable, i.e. without being in the beholder's eyes. Third, value can always be interpreted as instrumental: we do not just value things, we value them for *reasons*, thus making the valued act, object or situation instrumental to the satisfaction of our reasons. The one reason left, then, for attributing intrinsic value to anything cannot apply to nature. If value is seen as instrumental, as a property of a means to an end, there must be either an ultimate end or an infinite regression of ends that are valuable as means to further ends. We assumed the latter to be undesirable and remarked that the 'ultimate end' does not have to be an intrinsically valuable end. Still, if it were, it would have to be something inherent to the valuer rather than 'nature' – if nature or any property of nature were intrinsically valuable, any part of it would be valuable in the same sense and it would be impossible to rank the parts. We would, for instance, not be able to choose between saving the antelope and harming the lion's interests.

There is only a very weak sense in which intrinsic value can nonetheless be attributed to nature: if we understand value as external, even instrumental, we may still be unable to indicate why precisely we feel that nature, naturalness or a natural object has value. We could call this intrinsic value by default, but a possibly momentary and contingent inability to explain our sense of appreciation in no way supports any stronger conception of intrinsic value. The analytical and practical problems involved in attempts to apply 'intrinsic value' to nature are one set of reasons why liberalism would have to avoid this option and stick to the idea that value is external or instrumental and that it exists only in the eye of the beholder. (As from now, I shall use 'external' and 'instrumental' as synonyms unless explicitly noted otherwise; in the latter case, external value refers to instrumental value plus all other unexplained or unspecified subjective reasons to attribute value.) In this section, I discuss a second set of reasons against intrinsic value: the precepts of liberal democracy itself. Our discussion of these conditions will lead us to the paradoxical conclusion that liberal democracy must be committed to an instrumental view of nature, that it can nevertheless respect the point of view of those who attribute intrinsic value to nature to a large degree, and that it can, in principle, produce the results that the latter would want a liberal democracy to produce.

The most distinctive feature of liberal democracy is its insistence on the liberty of life: it recognizes no higher standard of the good than that of the individuals' plans of life. In this respect, more substantive political theories are better off: either the individuals' preferences fit their standard and are accepted, or they do not and will be rejected. Hence, if we want to 'green' an existing substantive doctrine, we already have a definite standard of right and wrong. Green ideas will fit or they will not. Or, the other way around, if we want to extend and enrich a green political theory with elements of an established substantive theory, the latter will fit or it will not. Substantive political theories at least offer straightforward, clear and distinct standards of value. Liberalism does not. Being, at least in principle, open to the pursuit of all plans of life, all theories of the good and all standards of value, liberalism makes it far more difficult to determine which ideas and preferences (green or otherwise) will fit and which will not: there is no straightjacket to fit them into.

Initially, at the level of private convictions, beliefs and plans of life, liberalism accepts all preferences, all values and standards, as equal and equally respectable; it has no judgement on good and evil. At a collective level, though, where cooperation is at stake, this idea of equality is potentially hazardous. Social cooperation can only be guaranteed in full if individuals have identical or perfectly compatible plans of life and preferences. If the plan, values and theory of the good of just one person is compatible with those of others, their preferences may still be incompatible. As the then love of my life once wrote to me, you can't always get what you want. It is at this point that second-best, third-best or *n*-th best solutions to the problem of social cooperation have to be found, and it is here that liberal democratic procedures for collective decision-making must enter the stage. A kind of social preference ordering has to be achieved, which requires a choice and selection procedure, which in turn presumes a normative standard indicating why one solution is (*n*-th) *best* and another not. This is the liberal equivalent of a standard for the collective or common good.

There are, in principle, three possible sources for such a standard:

(1) an *independent substantive standard*, as a consequence of which individual preferences *can* only matter in so far as they deselect less desirable alternatives from among a set of 'acceptable' worlds, the latter being determined on the basis of the standard and independent of individual convictions. This assumes that the standard is still indeterminate enough to allow more than one possible world to qualify as good, and it assumes that individuals can veto alternatives. If not, then individual plans of life will be totally instead of nearly irrelevant.

(2) a *procedural standard* like (qualified) majoritarianism, selection by lot (cf. Goodwin 1992), the Borda rule, a measure for the highest compatibility of preferences, etc. The role of individual preferences in this is to limit the set of possible worlds to those at least one person finds at least minimally desirable; the outcome of the procedure need, however, bear no further relation to people's preferences.

(3) an *intersubjective standard*, i.e. one on which all those involved can or reasonably should agree. Liberalism would want this standard to be not just contingently intersubjective but principally *impartial*: that is, it should not only be acceptable for the present set of individuals with their present preferences, but in general be compatible with all the possible elements of the set of reasonable plans of life. Hence, this type of standard will be based on a hypothetical rather than an actual agreement (cf. the contract theories of John Rawls (1972, 1993), Bruce Ackerman (1980); Brian Barry (1989, 1995a)).

Alternatives (1) and (2) are not open to liberalism. A substantive standard is simply incompatible with the liberal rejection of moral objectivism. Note, however, that differences between choice *procedures* tell us little about the results: it is, for example, possible that a theocratic and a liberal democratic society opt for the same rules on the treatment of animals.

Procedural standards must be rejected for two reasons. First, because they need further justification (cf. Section 1.4) – either in terms of a substantive standard (which we have just rejected), or a procedural standard (*regressus ad infinitum*), or an intersubjective standard. Second, because from a liberal point of view there are things that are just too important to leave unprotected against the fad of the day no matter how small the risks are. One of the paradoxes of democracy is that it can be used to abolish itself; if given an unlimited scope in a liberal democratic society, it could be used to abolish liberty, equality and justice. Hence, procedural standards can only play a role in decisions on non-crucial, non-constitutional issues.

Having excluded the impossible, we are left with the option of an impartial standard only. I do not want to discuss the exact shape of this impartial standard here (there is more on this in the next chapter) but I do want to make some general observations. First and obviously, the standard will have to be justified in relative independence from current individual opinions, meaning (among other things) that even if all of humanity today felt like setting the planet on fire and going out with a bang, this in itself would not constitute a legitimate, impartial reason for a liberal democracy to decide accordingly.

Next, there is no logical reason why an impartial standard should necessarily be different from a substantive standard. It is imaginable that the reasoning for both would lead to the same standard for social decisions. Moreover, we saw that procedural standards can also be compatible with an impartial one, subject to respect for the 'things too important to risk'. An impartial standard may make everyone reach the same level of sub-optimality: that is, make everyone equally miserable in comparison to their individual dreams of a first-best world, but it can also result in preferential treatment for some individuals and plans of life to the (relative) disadvantage of others. Nazis, religious fanatics and rapists will most certainly be worse off than their would-be victims. If the precepts of liberal democracy imply care for nature, the same will be true for people with environmentally hostile plans of life. Note that this is a kind of unequal treatment that can be justified even, or especially, in liberal terms: if principles of justice chosen under conditions of impartiality work out badly for some, this is, by definition, because their plans of life were incompatible with the aim of enabling a maximum liberty of life for all. Those who feel they lose under such an arrangement would still have to accept its results, assuming that they are or as reasonable creatures should be committed to the ideals of liberal democracy (cf. Barry 1995a: 150).

We only get out of an impartial decision procedure what we put into it, i.e. if it is based on the overall importance of equality, liberty and the pursuit of a plan of life, then that is precisely what it will protect and nothing else. Our stringent conditions of liberal democracy do not, in rational choice terms, guarantee a complete social ordering. They do not preclude the possibility that our impartial standard will be indeterminate on some issues. In fact, that is exactly what a liberal democrat would want an impartial standard to do: to prescribe moral side-constraints to the free pursuit by individuals of their plan of life. In short, our impartial standard will create three spheres of decision-making (cf. Barry 1995a: 110, 149 ff.):

(1) *constitutional*: decision-making on fundamental issues, that is, when the compatibility of policies with the precepts of liberal democracy has to be established

(2) *political*: decision-making on issues that are politically but not constitutionally relevant, i.e., where and in so far as there is a choice between policies that are all compatible with the conditions of liberal democracy

(3) *private*: decision-making on issues that do not change the structure of society or social cooperation. In the words of Melanchton, these issues should be considered *adiaphora*, indifferent to salvation – or politics, as in our case.

An arrangement like this implies that some rights can be absolute, irrevocable, inalienable and so forth, whereas others will or can be (far more) conditional. This purely formal description of a liberal democratic system of 'social valuing' does not say anything definite about precisely which rights are and which are not absolute (we discuss this topic in detail in the next chapter). Nevertheless, we can make an important observation *in re* the protection of nature. Being instrument-ally valuable, nature can give rise to issues fit for democratic decision-making, both as a valued and as a disvalued object of individual rights. Yet at this point constitutional protection, and thereby protection 'as if' it had intrinsic value, is not excluded.

A further observation is that we can reconfirm the conclusion drawn in Chapter 2 on the way value is attributed in liberal democracy. An impartial standard of value conforms to the liberal democratic demand for equality and for treating the interests of all recipients as equally important. Unequal treatment requires a 'mor-ally relevant difference' between recipients. If nature, or rather parts of it (animals), can be considered recipients, as I shall confirm in Section 4.4, then they deserve protection as subjects rather than objects of rights. More importantly, the onus of proof for the legitimacy of unequal treatment will then rest with those who argue against equality.

Although the individuals' plans of life and their reasons for apportioning value are treated as given in *political* decision-making, this does not put them beyond *public* debate and scrutiny. In fact, there are good reasons to assume that a public debate on private convictions and preferences is useful. First, if the political system is based on an impartial hypothetical agreement among reasonable beings, it can only become viable if real people continually or at least occasionally 're-enact' the drawing up of the contract; hence, public debate in a deliberative democracy could be a necessary (though not sufficient) condition for the viability of liberal democracy. Moreover, public debate may enhance stability. In an optimistic mood, we would expect a debate on people's reasons and values to lead to consensus, mutual understanding, acceptance and toleration, not to mention an improved insight into our own convictions. Then again, the result might just as well be con-fusion, rejection, intolerance and discord. We return to this topic in Chapter 9.

Related to this is the discovery that whereas in public debate all types of value from intrinsic to instrumental and all sources of value from divine inspiration to pure convention (cf. Section 2.3) can be admitted and discussed, decision-making in the constitutional and political sphere will tend to favour preferences based on intuition or reason and will be based on considerations of instrumental value only. Liberal democratic political institutions are designed to transform individual preferences into rights in such a way as to reconcile the plans of life of individuals as best as possible. Hence, decision-makers are expected to consider the interests of *individuals*, including typically individualized sources of judgement like reason and intuition, rather than the desires of divine creatures or, in the case of conventions, the sheer absence of any considered reason. Furthermore, their concern with the reconciliation of individual plans of life compels decision-makers to put every issue, act and object in that perspective, i.e. as means to those ends.

A deeper reason why liberal democracy is bound to translate all individual preferences, reasons and values in instrumental terms is that it is concerned with individual liberty and autonomy. If a political system is to offer room for both, room for voluntary behaviour and moral rather than predestined choices, it must consider plans of life as ends in themselves, preference satisfaction as a vehicle, and the objects of preferences as means. Hence, even an individual's desire to see the intrinsic value of nature respected by others is, from the collective point of view, an end to which the recognition of intrinsic value is a mere instrument.

We should also observe that the political system's disposition to interpret plans of life in terms of external, even instrumental value does not prejudice it on the question of a deontological or consequentialist ethic of political action. It does condemn liberal democracy to a deontological as opposed to a teleological position (cf. the distinction made by Sandel (1982)): the system sketched here makes political institutions protect rights related to undefined plans of life, rather than develop and implement policies aimed at some common or highest good (*telos*). Yet the distinction between a political and a constitutional sphere leaves room for both deontology and consequentialism. Rights that are subject to constitutional protection will be beyond consequentialistic arithmetics. They define the polity's categorical imperative: that is, they describe what the polity should (not) do to act as if it were the law of nature. Rights subject to political decision-making, on the other hand, are typically open to (collective and individual) arithmetics.

Finally, we can make an important remark on the possibility of offering nature constitutional protection. In the course of the real-world political debate on the greening of liberal democracy, the issue of the constitutional status of the environment and of environmental policy is bound to pop up. Similar issues are already on the political agenda in many of its member states: animal rights in Great Britain, for instance, the intrinsic value of nature in Germany and The Netherlands. In view of what has been said above, we should note that a constitutional recognition of the intrinsic value of nature, at any rate, will be nearly meaningless except as a political move. It may serve to gain support from environmentalists and it may inspire public deliberation on politics, but it cannot serve to inspire a liberal democratic green policy; the latter will necessarily be inspired by instrumental, in a sense anthropocentric considerations. The same is essentially true for the issue of animal rights. Unless the basic tenet of liberal democracy, the sanctity of plans of life, is amended, there can be no guarantee, no positive constitutional protection, for the welfare or interests of any creatures other than humans. For deep ecologists, this may be a diabolical conclusion; for others, it may reek of *Realpolitik* – but even idealism cannot prosper without an informed judgement of reality.

4.3 Substitutability

From a liberal democratic point of view, we saw in the previous section, all value can and must be considered instrumental to the end of fulfilling a plan of life, even if an individual themselves believes in the intrinsic value of – in our case – nature. Liberal democracy is concerned with the conditions under which all individuals can pursue their best feasible and n-th best course in life. Against this background, it is imaginable that things which some individuals value could be transformed into things others (or they themselves) value more: specifically, that nature could be traded in for resources, machines, consumer goods. The question I discuss here is whether such a trade is more than imaginable – is it actually possible and permissible?

In a nutshell, the question is this: can alternatives have the same value as originals? Can one 'vehicle of value', say, an old chestnut tree, be replaced by others, say, a painting or hologram of the tree in a museum, a load of furniture, a series of poplars, a house, cash, or any combination? Any answer to this question can have extraordinary consequences. If there can be literally equivalent substitutes for nature, not to mention better alternatives, we have a *carte blanche* to get rid of all these unorderly shapen, inefficient, bug-infested trees, plants, animals and rocks. If substitutability is impossible, each single part of nature would, for all practical purposes, have a unique kind of intrinsic value that would make all talk of value absurd. It would then be impossible to distinguish between the rightness or goodness of turning a raw carrot into something edible, or to judge the relative goodness or rightness of the carrot-into-food transformation on the one hand and nuking the Amazon rainforest on the other. Finally, if substitution is possible only within limits, the exploitation of nature may require that (necessarily inadequate) replacements be found, or that compensation be made available – but what exactly can compensate for the loss of a tree?

Received wisdom allows us to distinguish three main notions of substitutability, to which I want to add a fourth. The two most classical notions are weak and strong substitutability, both of which are fundamental ingredients of two distinct (respectively strong and weak) interpretations of sustainability (cf. Beckerman 1994, 1995; Holland 1996). Sustainability relates to how much is and can be consumed, while substitutability is concerned with the kinds of goods that can be consumed (cf. Common 1996). Strong substitutability (linked to weak sustainability) means that natural and artificial capital are interchangeable, i.e. that (1) any natural good can be replaced by an artificial good if both result in the same level of welfare; and (2) all natural goods can in fact be replaced by artificial alternatives. Proponents of weak substitutability (strong sustainability) accept (1) but reject (2).

In other words, its advocates assert that there are things in or aspects of nature that cannot possibly be replaced by anything man-made (cf. Holland 1996).

The term welfare, as used in these descriptions, must not be interpreted too economically in financial terms only. It can easily be replaced by value in general. In calculating the welfare we derive from nature we must, as Wilfred Beckerman remarks, take account of all the factors that contribute to welfare: '. . . environmental, distributional, social and other considerations' (Beckerman 1994: 200). The idea is that welfare is a cumulative measure for all the values we attach, for whatever reason, to goods or situations. Hence, the one argument that cannot be used against a conception of strong substitutability, whether in terms of value or in terms of welfare, is that it does not take account of any particular factor, say, beauty or health.

Yet there is a powerful argument against the distinction between weak *and* strong substitutability, to the effect that it can be reduced to strong substitutability. Strong substitutability accepts that natural capital is replaced by artificial *provided* that there is no decline in welfare (Holland 1996). Weak substitutability:

> asserts that there are significant elements of natural capital for which no human-made substitutes can be found . . . [which] . . . amounts to the claim that nothing humanmade can provide mankind with the same degree and/or kind of benefits as those provided by the elements of natural capital in question. But if this is true, then the loss of such natural capital would inevitably mean a decline in human welfare and would not therefore be counteracted by [strong substitutability].
> (Holland 1996)

Holland then concludes that nothing which is counteracted by weak substitutability is not also counteracted by strong substitutability: the idea that the two 'differ in any important respect turns out to be a charade' (Holland 1996).

A third conception of substitutability is associated with 'absurdly strong sustainability' (Daly 1995; Holland 1996). I shall refer to it as absurdly weak substitutability. Its proponents, Daly and Holland, argue for a moralized interpretation of substitutability: natural capital should not be substituted by anything humanmade *even* if it is possible. Holland (1996) points to two types of argument in support of this position: non-human interests and the independent value of nature.

Holland argues that the moral claims of at the very least sentient animals put constraints on our desires to transform natural into (only for humans useful) artificial capital, thus in a way extending weak substitutability to include the welfare of non-human beings. But he also feels it must be possible to move beyond all consideration of welfare by recognizing that 'the natural world has a special significance and importance in our lives' independent of its being good or bad for us (Holland 1996). This special significance seems to originate in nature's being a necessary condition for our very existence and/or in the origins of nature itself – its being 'original' rather than humanmade (cf. Goodin 1992). Holland admits that this provides insufficient grounds for us humans not to 'do what we like with the little bit of it which happens to lie close at hand' because we have a whole universe to look upon as significant, meaningful, context-giving and original – but the point of the argument is that it gives us a reason to value the continuation of the natural world *as such*. Thus, as he writes elsewhere, *parts* of nature can be substituted by artificial capital, but the whole has to 'retain ... something of its original form and something of its identity' (Holland 1994: 178): there are 'units of significance' that cannot be substituted by anything humanmade. In the terminology of economists, Daly's and Holland's claim is that perhaps every part of nature can be exchanged for something artificial with the same value – except for its quality of being natural, original, the product of natural history. There is no way a human can reproduce that, whereas it is something we derive welfare from.

The problem with Holland's units of significance is not that he requires some conception of intrinsic value to make sense of it; he does not. Units of significance can be represented as being in the eye of the beholder, as being valuable because of the meaning imported on nature by the valuer, and as valued as instruments to experiencing originality. Nor does Holland make the popular mistake of arguing that welfare cannot represent all we value about things. His argument is that there is a type of good that cannot possibly be replaced by anything else, however valuable for whatever other reasons, without losing one specific value, one specific source of welfare. The problem with Holland's argument is not its weakness but its very strength.

If the originality of, say, a tree or a forest cannot be reproduced in the form of any imitation of the original, this has two reasons: because we value originality, and because originality as a source of value cannot be traded in for anything with another source of value. The catch is in both parts.

To start with the latter reason: if one source of value, one type of value-adding property, cannot be exchanged for another, we must expect this to apply to properties other than originality as well. Let us exchange the *Mona Lisa* for a perfect copy; the original ceases to exist. The knowledge that it is a copy we are now looking at will take away at least one reason to value it, thus at least part of its total value and part of the welfare it produces. Now let us exchange the *Mona Lisa* for *Guernica*. Both are original, yet we value them differently because of other properties – colour, composition, style, subject, history. The *Mona Lisa* and *Guernica* can perhaps produce the same *total* welfare but not the same *kind* of welfare. It is the composition in terms of (valued) properties of any particular object that gives it a *specific*, unique quality that cannot in any way be reproduced, exactly because it is unique. What goes for the originality and natural history of natural objects goes for any object: they are valued because of their properties and the unique mix of properties that make them distinct objects.

Catch number two is in the 'we' part of 'we value originality'. As John Barry tried to point out to me, both 'we' and our notion of what is original are social constructions; we might well value the originality of a newly planted forest, even a park, rather than the non-originality of nature repeating its old game with every other tree and forest. But even more important, 'we' do not value originality – only some of us do, and even then not all the time. It is a private affair. Personally,

I do feel some sense of loss and defeat when I gain a decisive victory in the battle for sleep over a gnat, but I have no reason to assume this feeling to be universal. Some of us appreciate the *Mona Lisa*, some *Guernica*, some both, and some would not notice the difference or care. For me, my grandparents' chair is irreplaceable because of its history; for you, it is probably just a cheap, ugly, old and unpractical thing for which millions of substitutes exist, all superior with regard to any of the properties you care about.

What we have here is a fourth conception of substitutability: warping outrageously weak substitutability. It admits that natural and artificial capital can be substituted for one another without change in the level of welfare, yet is impossible if the value-adding properties of two objects result in a different composition of the welfare each produces for any particular individual. And just to make life difficult, this is precisely the view on substitutability that is best compatible with liberal democracy, since liberal democracy does not care in the first place about overall welfare but about the welfare of the individual and the individual's perception of value. This is not to say that substitutability is *in principle* impossible; it all depends on the individual's theory of the good, on their perception of value. If you can replace your neighbour's dead goldfish by one that looks alike, and they do not notice, it was substitutable even in the 'normal' weak sense, at least for them. If you could replace it with a Korean electronic goldfish, and they would not notice or care if they did, the goldfish was substitutable even in the strong sense. Thus, substitutability is not impossible – just close to impossible.

We have reached a perplexing conclusion: liberalism seems to have a potential for being more radically green than any radical green theory we know of. It is possible for a liberal to argue that nature and elements of it should be protected against human interference even if we attach only instrumental value to nature. Yet the price we have to pay is tremendous: we must accept that nature cannot be substituted by artifice. As long as there is one individual around who values nature for its being natural, there can be no simple trade-offs: someone is bound to lose. Unless we get help from outside, develop new arguments, substitution is not just practically impossible but also prima facie impermissible. I shall let this conclusion simmer for a while and return to the question of legitimate acquisition and use of nature in the next chapter. I first want to complicate matters further by arguing for another unsettling conclusion: that there are morally relevant subjects other than humans for liberal democracy to reckon with.

4.4 Inclusion and exclusion

The last question I discuss in this chapter concerns the rights of animals, plants and rocks – or in more precise terms, the status of non-human nature as a morally relevant subject rather than object. There are two reasons why we should now turn to this issue. First, I have in this chapter (most of the time) carefully avoided references to humans as the sole source of value in the universe; instead, I spoke about persons, individuals or valuers. In view of the fact that at least some animals are said to be valuers in a way (cf. e.g. Agar 1995) and that their interests therefore deserve to be taken as seriously as those of humans, we have reason to ask whose interests, values and plans should matter to liberals – and in which way. Moreover, regardless of whether elements of nature are substitutable or not, a view of nature that precludes recognition of the possibly intrinsic value of nature precludes prima facie protection of nature on the basis of its moral status as an object, but not necessarily as a subject. That I value my home help as irreplaceable tells us something about how I value her as an instrument to the perfection of my home, but it does not follow that I am permitted to lock her up in a cupboard or rent her out to others if I so desire. It is customary, in liberal theory, to grant humans an immunity against subjection by others – though not necessarily because of their intrinsic value, as we shall see. Whatever they be, if we have reasons to limit the use of other humans as mere instruments and if there is no reason why they should apply to humans exclusively, we have to ask ourselves which other creatures they might protect – say, angels, Martians, or animals. Hence, the possibility that elements of nature can be subjects in their own right may offer them, within a liberal democratic world, a kind of moral protection that insistence on their intrinsic value as objects cannot give them.

The arguments in this section presuppose that a subject, apart from being an entity endowed with certain properties, is an object to which certain rights have been attributed. If we assume for a moment that having a plan of life is what makes a being a subject, then my home help, for instance, is a human being endowed with a plan of life and an object to which the right of self-determination should be attributed. The language of rights used here will not be explained (nor need it be explained) until the next chapter. Nevertheless, it is appropriate to say that the class of rights discussed in this section is that of what I call *inalienable* rights, i.e. rights that can be recognized and in this moral sense 'attributed' to an object, but that cannot legitimately be 'distributed', denied or alienated because they exist outside the realm of politics and distribution.

The issue we face (cf. Section 2.3) is that of a choice between attributing the status of subject to humans only (anthropocentrism), to intelligent creatures (intellectualism), to sentient creatures (pathocentrism), to animals (zoocentrism), to life (biocentrism), to nature as a whole (ecocentrism) or to let our judgement depend on the development of our morality (evolutionary ethics). I shall exclude evolutionary ethics on purely analytical grounds later on; I first want to focus on the question whether liberalism can 'extend the circle of compassion' (cf. Gold 1995) beyond humans. In an odd analogy with feminism, there seem to be two perspectives: a difference approach and a likeness approach.

The difference approach in environmental ethics (as exemplified by e.g. Cooper 1995) differs fundamentally from that in feminism: it is not concerned with differences between humans and other creatures in moral capacities, reasoning or concepts (cf. Kymlicka 1990), it is not even concerned with being different as such but only with being *the* different. Its central concept is Levinas's description of experiencing the Other, a subject, in place of the other, an object. A classic example is the murderer who looks into the eyes of his victim and discerns a presence, a consciousness other than his own, and with that a question that wants a response, an appeal for justification and responsibility, a criterion of morality that overrides his Self and its limited perspective. A similar approach in environmental ethics would have us look at nature or elements of it as our Other, questioning our unconscious motives, provoking us to justify our actions intersubjectively.

The problem with the difference approach is that it is not too different from what we would want the likeness approach to be. Both assume the existence of the same (initial) point of reference: the Self. Both use the eye of the beholder to perceive and recognize the moral import of others. Consequently, both invite the valuers to acknowledge the limited nature of their own perspective and the prima facie validity of the moral perspective and claims of others. At the very least, then, the difference and likeness approaches are compatible. The disadvantage of a difference approach is that it fails to give reasons for recognizing an other as Other; the advantage of the likeness approach is that it does, or at least tries to.

The more popular likeness approach is followed by authors ranging from Peter Singer (1977), a utilitarian, to Tom Regan (1984), who takes an opposite (deontologic) position by arguing for individual rights for animals. The basic unit for these and other writers is the Self: there is, they assume, something about myself that makes me morally important. It would be a purely subjectivistic, even solipsistic, approach if it did not immediately question this axiom and demand reasons: am I important because of my plan of life, because I am an end in itself, because of my opinions, because of other properties, and if so, which? Assuming these properties to be known, logic forces us to recognize other objects as equally worthy of moral concern if they have the same properties. From this point on, the likeness approach can develop in several directions. If we have reason to demand that other entities have to have all the properties I have, the combination of which makes me unique, it results in solipsism. If the criterion for relevancy is that others have (only) all the properties that make me important, it usually results in anthropocentrism. If it is that others should share some properties, it will result in an egalitarian recognition of all humans, all intelligent or sentient beings. Finally, if it allows recognition as a subject to the degree that properties are shares, it becomes a hierarchical ethic. It is this last version of the likeness approach that I defend here.

As a first step, let us consider the axiom that I matter. It is an intuitive axiom and one that is as such questionable – but less questionable than it might appear to be. Not believing in an afterlife, it seems odd to say that I matter if I would also sacrifice my life, a necessary condition for my existence, to catch a bullet for someone for whom I care. Apparently, I do not care enough about myself to care about my existence. However, I care about at least some others, which I would not be able to do if there was no I to value their existence. My existence, then, is a necessary condition for being able to care for the people or things I really care about. I may not care about my existence *as such*, but since I care about others I should care *instrumentally* about my existence. In more general terms: I care, therefore I matter. We can generalize this even further: as long as any Self cares about anything, values it, the Self matters.

The obvious objection to this conclusion would be to ask: what if I do not care about anything? There is a logical reply to this, i.e. that I still care about myself at least as much as I do about other things, but it cannot really take away the objection. The only sensible reply I can come up with is that any conclusions we draw from our axiom will be both as valid and as irrelevant as any other theory of value, and that all debate ends there. In fact, there would be no debate: if we really did not care about anything, we would not take the trouble of reading this.

Assuming, then, that I and the reader are still around, hence that the Self matters, the next question is why it matters. Now although this is a deep meta-ethical question, I cannot discuss it as thoroughly as it deserves. The aim of this book is to see if the precepts of liberal democracy are compatible with environmental concerns, not to question those precepts themselves. Hence, I must restrict the discussion to liberal reasons for the moral relevancy of the Self. Liberalism gives us two reasons: because man is an end in itself and because humans have plans of life. The first, the classic Kantian reply (cf. Kant 1928, 1974), can be reduced to the second, the reply of late twentieth-century liberalism.

If humans are ends in themselves, it is because they are, or rather can be, autonomous: that is, because they are free to choose the rule(s) or maxim(s) directing their lives. Were they not, they would live heteronomously, i.e. live under the law of others or something other. They would be slaves under the yoke of nature, not choosing their own ends but having their ends and their own role dictated by external forces. They would be merely means in a universe of means to ends that are means to further ends, in infinite regression. Humans being ends in themselves is the meta-ethical *ne plus ultra* for Kantian liberals: the Self matters because it must matter, because the only alternative would be a universe in which nothing matters.

Human autonomy's reflection in maxims is a special case of its being reflected in plans of life. A plan of life, as indicated earlier, consists of a theory of the good (our goals in life) plus further rules on how to live accordingly. These are what a Kantian would call maxims. The difference is that in the Kantian theory of the subject, autonomy requires a categorical imperative as the highest rule of all, the maxim for the selection of maxims – in the case of Kant himself, the rule that we should want our maxim(s) to be universalizable: that is, laws of nature or positive laws. The modern liberal conception of a plan of life allows but does not require a categorical imperative to be part of a theory of the good. It requires the *possibility* of autonomy, not its actual existence. In the absence of an ultimate ethical truth (see Chapter 1), it considers the demand that our maxims ought to be universalizable as either impossible or undesirable. It thus allows human life to be a project, a plan, in which the discovery of a categorical imperative can be something to strive for but not something we should expect to find.

We should note two things about the modern liberal perspective on autonomy. First, it is open to the objection that without a categorical imperative the choice of a plan of life can only be a free choice in a very limited sense. Ultimately, it seems to be determined by external factors – education, culture, ultimately nature. In an ironic way, this brings modern liberalism closer to nature than classic Kantian liberalism. Second, the notion of a plan of life with all its limitations is still the liberal *ne plus ultra* in the defence of humans as morally relevant subjects. Without even so much as the possibility of autonomy, nothing would matter, all talk of ethics, morality and politics would be pointless.

In liberalism, then, we must assume that having a plan of life is a necessary and sufficient condition for being considered a subject. Consequently only humans, extraterrestrial intelligent creatures and Kant's angels (Kant 1974) would qualify – no other element of nature could. However, a closer look at this condition will reveal that it is in fact made up of a set of conditions, all of which together are necessary conditions for calling a plan of life a plan of life, yet two of which are by themselves sufficient to qualify an element of nature as a subject.

To have a plan of life requires life, consciousness and agency – life being a necessary condition for the others but insufficient to make an object a subject. Let me first show that these are necessary conditions and then why the last two are sufficient.

I understand life here as a primitive notion: it is a particular kind of existence, different from being a character in a novel and different from being unable to change though internal mechanisms (a chair reacts to outside forces at the level of single atoms or molecules only, a tree also adapts, it reacts by internal mechanisms more complicated than single and distinct molecules). It further requires existence on a similar plane with us (you and me) in the sense that the laws of nature should not prevent us from having a reciprocal relation of contact and influence with it, a kind of society in a very loose sense of the word. I realize that this is an imperfect, conventional and perhaps arbitrary definition of life and that there is much more to be said on the subject, but this is not the right place and time. The point is that without life in this sense, we cannot treat an object x ethically nor can it treat us ethically: it cannot act and it cannot be conscious of its acts or of ours.[14]

To have a plan of life requires consciousness, not necessarily self-consciousness but merely consciousness. First, the rules of semantics make it impossible to refer to anything as a subject without ascribing conscience to it. If y is not conscious of anything – itself, surroundings or the distinction between both – it can be part of a plan or project but not have or implement one itself. More importantly, without consciousness y will be unable to value, that is, to distinguish damage and improvement from harm and benefit. And without this, the knowledge of subjective good and evil, there can be no ordering of alternatives, no preferences, no reason to act in any specific way – no plan.

To have a plan of life also requires agency, the ability to act either physically or mentally (design goals, intentions, attitudes) or both. A plan that cannot be executed is still a plan, an agent y who cannot execute a particular plan is still an agent, but an entity z that cannot ever execute any plan at all cannot have a plan – z is not an agent but passive.

Agency and consciousness together are required to make sense of the notion of a plan of life, but each separately is enough to qualify as a subject, even from a liberal perspective and despite the liberal insistence on the basicness of a plan of life. In environmental ethics (cf. Larrère 1997) and even in liberal theory this claim is often supported by an analysis of the moral status of physically or mentally challenged humans. A liberal will typically argue that we would not want to treat comatose persons like or worse than a chicken – that is, kill and eat them – even though a chicken has conscience of sorts but no agency and the comatose have neither. An animal rights' advocate will argue that if we must treat the comatose as ordinary human beings, we are even more obliged to treat the chicken like a human being. I would not want to argue my case this way, partly because I do not want to go this far and partly because the comparison between pigs or chicken and the physically or mentally incapacitated tends to be insulting for the latter and those who care for them.

Instead, I ask the reader to imagine ordinary human beings who, for reasons that are theirs and (as liberals) not ours to judge, become enchanted by the beauty of Buddhism or some radical version of stoicism. Rather than accidentally entering a comatose state, they choose to give up either every desire to act or all consciousness. In both cases, they no longer have a plan of life; their plan is finished, completed. In the first case, they are still conscious of themselves and things around them, they just do not (re)act. We know that we can still harm or benefit them rather than damage or improve them: we will still consider them a subject. In the second case, they act but are not conscious: as stoicism and virtue have become their second nature, they adapt their attitudes without thought. Again, we will still consider them a subject, since it is they who act and no one else.

If we believe that the mentally and physically handicapped should be treated as, not merely like, human beings, we do not do so because the former are 'possible' humans. It may be a quirk of nature or fortune that has made them different from us, but similar quirks make chicken, bacteria, trees and rocks different from us. We see them as subjects because the difference between them and us is not *relevant* and the similarity is. They differ from us and cannot possibly be like us, but as our stoic Buddhist shows, *we* can become like *them* and we would still feel that we matter.

There are two points to be made here – one very radical, the other extremely conservative. First, since it is relevant likeness that obliges us to recognize others as subjects, then if we value ourselves as subjects because of our properties of agency and consciousness, we should value other entities with the same properties in the same way, i.e. as subjects. Thus, if we have reason to believe that some animals are conscious, to a degree even agents (Agar 1995; Hargrove 1992;

Benton 1993), we should consider them as subjects and represent their interests in liberal democratic decision-making, full stop.

The second point is that recognition as a subject need not make all the difference in the world. We can appreciate two entities equally as subjects without being obliged to treat them equally. The difference in the capacity for agency between me and a comatose person does not make the latter less of a subject but the *degree* of difference in this very same respect can justify unequal treatment. No reasonable government will put a comatose person behind bars for not paying their taxes on time, yet this could happen to me if I am late. More subtly (although the example is rather crude), if I am deep asleep while the person next to me is strangled, I cannot in reason be held responsible for what happens; if I am wide awake and do not act, I make myself an accomplice; and if I am drunk or stoned, the assessment of my responsibility will at least in part depend on the degree of intoxication. By the same token, we no longer prosecute rats for munching away at our grain or cats for failing to kill rats. (Humans did in the good old days – cf. e.g. Tester (1991: 67).) In short, even though some animals may count as subjects, it is possible to rate their and our interests hierarchically. This does not imply that *all* human interests necessarily take precedence over *all* animal interests – that it would be legitimate for me to eradicate all zebras just because I do not like the Rietveld kind of abstract art on their skins. All it means is that human interests *can* be more important than those of animals – and vice versa (cf. Sterba 1996).

In terms of the language of formal rights (see Section 1.3), what we have seen is that there are objects that, by virtue of the properties of life, consciousness and agency, have the special status of subject. A subject is an object endowed with certain inalienable rights, rights that cannot be denied to it.[15] However, what may be an inalienable right for some subjects (objects) need not be inalienable for another subject (object). Since I shall not be discussing animal rights too often after this, let me (very briefly and superficially) discuss vegetarianism, i.e. the hypothetical obligation for humans not to kill and eat animals or, alternatively, the rights of animals not to be killed or eaten.

In the case of vegetarianism, there are at least three issues at stake. One is whether or not it is permissible to eat (dead) meat, the second whether it is permissible to kill animals, and only in third place do we have to face the question whether it is permissible to kill for food. We cannot ascribe the status of subject to dead meat any more than to a brick wall or even a living vegetable. Whatever rights dead meat has are rights we grant, rights that are ours to give. Hence, there is no objection out of principle against us eating dead meat. I have heard people argue that we should not eat dead animals any more than dead humans, but this only establishes that if we cannot eat animals we cannot eat humans and if we can eat animals we can eat humans. There may be a taboo on cannibalism, but 'taboo' is not necessarily shorthand for 'good reason', no matter how much I myself would dislike breaking the taboo.

Are we permitted to kill animals, other than in cases of self-defence? On the one hand, animals are subjects, endowed with life, consciousness and agency. On the other hand, they do not have the same degree of consciousness, for all we know. Whereas killing a human being means that we end a consciousness *and* plan of life, killing an animal – again, for all we know – only means ending a consciousness. An animal can die from all sorts of causes, some more and some less painful, but it will not miss out on anything; it does not usually care about its future (it may care about its offspring but that is no proof of its caring about time; *pace* Benton 1993), and in that respect killing it does not harm it. What may make killing an animal harm is, first, the pain involved in dying, but if we kill it as painlessly as possible, as painlessly as nature allows, this objection disappears. Second, what may make it harm is the fact that *we*, *voluntarily* kill it – that we do harm. In different circumstances, we can do harm to others, voluntarily yet legitimately: we can punish criminals, we oblige drivers to stop in front of red lights – even closing the front door harms those who might otherwise have wanted to enter our house and have a look around. All this is or can be harm, provided we frustrate others in the pursuit of their plan of life. The important and for vegetarians miserable thing to note is that the permissibility of killing animals is again not a matter of principle: if our reasons are good enough, we may kill them. And, with the same qualification, eat them.

What I have said obviously applies only to some arguments in favour of vegetarianism. There is far more to be said on the subject, some of which may possibly give sufficient reason in support of vegetarianism (e.g. the environmental costs and inefficiency of cattle farming as compared to growing vegetables), some of which applies to humans as well (the sanctity of life as a prima facie reason against killing) and some of which would be out of place (the conditions under which animals are raised bears no relation to whether we may eat them, only to how we should treat them). I would not in any way want to claim that vegetarianism is based on fallacies or misguided ideas or feelings. I would want to say this, though: it seems that it does not really matter if animals (or angels, for that matter) are recognized as subjects; it will probably not protect them against being eaten. We may therefore, perhaps surprisingly, expect that their having *instrumental* value, value as objects, to humans can offer them at least as much protection as the status of subject. This gives us all the more reason to turn to the next chapter and ask how we are to treat *objects* of value – man, animal, plant, mineral.

Having discussed ethical theories based on properties, from anthropocentrism to ecocentrism, I want to end this chapter with a few words on the one remaining position: a theory in which moral relevancy depends on relations instead. According to the precepts of evolutionary ethics, we should care about objects other than ourselves as much as we have reason to and to the degree that our moral sensitivity to non-human nature develops in the course of our cultural evolution (see Section 2.3). There is as little reason to reject this view as there is to accept it: as a descriptive meta-ethical theory, it is a truism; as a prescriptive ethical theory, it confuses fact and value, context of discovery and context of justification. If we *cannot*, at time *t*, recognize, say, animals as subjects, then we are excused; if we can, if we have the necessary knowledge, culture, intellectual environment, then we should. In this respect, evolutionary ethics adds nothing to existing views, except perhaps the undefended idea that there is a 'right' ethic for every stage in the development of society. If evolutionary ethics were used to justify an extension of the circle of interest (cf. Bosselmann 1992; Wenz 1988), then the only justification it offers is a kind of positivistic identification of morality with current opinions, which prevents it from saying anything about how we should behave in the future. If it were used, finally, to justify our present ethics (assuming there was only one system of values in operation), then it would reduce the goodness or rightness of an ethic to its being appropriate to time and place, i.e. it would reduce ethics to historical necessity. And if we believe in the least bit of freedom of choice, or alternatively believe in the absence of guidance by Fate even though we believe freedom of choice to be a deception, then we must also believe in ethics and reject necessity.

5

The distribution of rights

5.1 Spheres of rights

The leading question in environmental ethics is: what are we permitted to do to nature? The big question in environmental politics is: how are we permitted to do it? Both questions divide the world into things we can and cannot touch and acts we should and should not perform. They divide the world into rights.

In Chapter 1, I described the central place and role of the notion of formal rights in liberal democracies. Rights are, by definition, the subject of *attribution*: even if there were such a thing as natural rights, their existence would still have to be recognized in the 'real' world of law and morality. We, as real people, must at least know of their existence and have reason to affirm them, or else they would be meaningless and as arbitrary as the results of a lottery. In a loose sense of the word, attribution implies *distribution*: affirming that I have a right to produce plastics comes down to the same thing as taking two particular sets of molecular rights (cf. Section 1.3) out of the set of all rights, handing one of these sets over to me (production rights) and another to others (duties not to interfere with my rights). I intend to use the term distribution in a more strict sense, though.

In this chapter, I argue in favour of a distinction between three types or spheres of right: inalienable rights, unconditional rights and conditional rights (Section 5.1). The last two can be both attributed and distributed, the first can only attributed – inalienable rights are not part of some social stock owned by no one or everyone. Inalienable rights are the rights that make a subject out of an object; unconditional rights are rights that subjects need to meet the basic standards of their plan of life (or an equivalent thereof); conditional rights are rights to everything that goes beyond this category of basic needs.

Rights in the everyday sense (complex rights, cf. Section 1.3) are not trumps or shibboleths; they, or at least most of them, are not sacred. We can imagine a world where all rights are absolute (valid everywhere and any time) and inalienable (non-transferable). In practice, however, a world of merely absolute and inalienable rights is impossible to live in and incompatible with liberal democracy. It follows that the spheres of conditional and unconditional rights are not empty. At least some rights will be limited in time, place, means, purpose or mode of performance: some rights can legitimately be transferred (alienated), some can be both alienated and attributed only under conditions. In both cases, it is the better argument (in terms of our criteria of liberal democracy) that determines whether a right should be attributed to subject a rather than subject b.

The criteria for attributing inalienable and unconditional rights are, as it turns out, relatively clear-cut. A liberal democratic political system has certain duties towards its subjects – put differently, subjects have certain rights – that justify the exploitation of nature. To put it bluntly, subjects have a right to eat when they starve, albeit the last food on Earth. What can be less clear, though, is *how* rights of this sort should be exercised. If our starving subject has a choice between eating the last dodo and dining on a far from extinct race of vegetables, we enter a more controversial area: that of conditional rights, of rights to distinct acts or goods that are, individually, not necessary to survival or the successful pursuit of a plan of life.

In subsequent sections, I therefore focus on the sphere of conditional rights. In Section 5.2 I introduce the restraint principle, a special condition on the distribution of rights which demands that rights to (in a physical sense) scarce goods be distributed in such a way that they remain, within the limits of necessity, available for further use. Hence, for example, no tree should be cut down unless there is no alternative; if it is cut down it should be replaced by an equivalent unless there is no alternative; and if it cannot be replaced, a proper compensation should be made available. The arguments on the basis of which I defend the restraint principle are almost purely formal: assuming that x and y are objects of conditional rights, no reason to do x is good enough to justify any further act y that is not necessary to x, and no argument in favour of doing x can ever be final. Thus, for example, burning down a farm to roast a pig is not as a rule necessary to the purpose, and my hunger may be insufficient reason to roast this particular pig or any pig. As a consequence, the restraint principle puts the onus of proof for every conditional right to exploit nature on the exploiters: nature is not up for grabs but should be protected against destruction – as much as possible.

The restraint principle is not only compatible with, or even a logical implication of, the idea of conditional rights. It is also an acceptable representation of what I shall demonstrate to be a necessary condition for the survival of liberal democracy: a reasonable scheme of cooperation between coexisting generations (Section 5.2). With a little help from John Rawls, I hope to show that a realistic view of (human) society must take account of the fact that the benefits of social cooperation go beyond the creation of a 'surplus' relative to what individuals would produce in a fictitious state of nature. Humans of *different* generations depend on one another, and the fact that existing generations have to cooperate is sufficient to justify a – what Rawls called – savings principle for justice between generations. The restraint principle is precisely such a principle – and it turns out to be

116

compatible with Rawls's vision of justice between generations as well as any other liberal view on that matter. In ensuring justice between present generations, the restraint principle also implicitly yet effectively answers the green call for justice between present and future generations.

Before we can even begin to answer any question relating to the attribution of rights, we have to know *who* recognizes and attributes rights – that is, whether *we* are or can be qualified to discuss this. Rights do not fall from the skies, not even if adjectives like 'natural' or 'divine' are attached to them. They have to be recognized or else remain pure fiction, and recognition is the prerogative of human beings, of the subjects of rights. Only those who can follow a law can give it, only those who can be moral can design criteria of morality.

This is not to say, first, that rights, laws or norms cannot ultimately have a natural or divine origin – but unless we ourselves have some reason to believe that a specific norm should be accepted, we cannot be said to have genuinely accepted it. Assume that the eleventh commandment had been: 'Thou shalt never say never.' We do have reason to believe that this should not be taken literally, or else the mere act of pronouncing this commandment would be a violation of it. (Besides, Alfred Russell has specifically forbidden us to see sense in self-referential statements.) Yet it would still be a meaningless rule as long as we have no reason to believe that, say, promises or commitments either cannot be formulated clearly and precisely or that they should not contain the term 'never'. In the first case, a promise 'to do *x*' becomes a promise 'to try to do *x*', abbreviated as a promise 'to do *y*', which is a promise 'to try to do *y*'. The commandment then translates as a prohibition on promises: 'never make a promise'. This obviously gets us nowhere. Alternatively, our eleventh commandment orders us to formulate promises in ways that avoid the term 'never', but it allows alternative constructions using 'always' and 'not' – in which case the commandment does not really make a difference. In the end, then, the eleventh commandment, taken on its own, is a silly rule. Either we find reasons to accept it, and then these reasons are what make us accept the rule, or we accept it on authority, in which case we accept authority and not the rule itself. Note that a similar objection applies to liberal democracies imposing laws on its citizens: they can force an individual into subjecting to a specific norm or distribution of rights but not into accepting it.

Saying that only those who can follow a law can give it does not imply, second, that only humans can ever create or accept norms. We have no reason to believe that we can never communicate with extraterrestrials or that apes can never evolve to our level of intelligence – we just cannot communicate with either one and will have to manage on our own.

The point of all this is that norms and laws cannot be accepted, nor rights attributed, without convincing arguments. This makes everyone who is able to distinguish a good argument qualified to discuss the attribution of rights – but what is a good argument? At this point, I have to make a relatively arbitrary move. Although there is much more to be said about this, given the context of this book

I shall simply assume that we already have a standard for a convincing argument: our set of criteria of liberal democracy. A defence of the claim that these are reasonable criteria for a just society, that reasonable and impartial persons should accept them, and that the acceptance of norms requires reasonable and impartial persons will have to wait for another occasion.[16]

Now that we have established our mutual qualifications, we can get down to business. As a first step towards establishing that there are three spheres of right, I shall reject the idea that all rights must be absolute and inalienable. Let us assume that there is always a material aspect to rights. Attributing rights invariably implies attributing rights to *goods*, to material things: if you have a right, you own a good. Without tools and resources, we cannot create products; without a tongue, we cannot speak; without the right to use one's brains, even the right to freedom of thought becomes void. Hence, what must be proven is that *either* (1) there need be no such thing as legitimate original acquisition: that is, a right to take goods (apples as well as brains) from nature, *or* (2) that there are goods that can subsequently be transferred to others (i.e. alienated) – *and* that rights to goods need not be the absolute rights commonly associated with property.[17] Once this has been done, we can not only distinguish between inalienable, unconditional and conditional rights, but we can also specify which rights belong in which category for which reasons.

For a right to be inalienable, its object must first be possessed: that is, legitimately acquired from nature. This raises a classic liberal question: the legitimacy of original acquisition. The equally classic liberal answer (i.e. John Locke's) assumes that individuals can acquire goods because they have initial rights associated with 'self-ownership': survival of the self can justify the acquisition of food, mixing the self with nature by means of labour justifies possession of the harvest, monopolizing more of nature than survival requires, thus depriving others of the means to survival is illegitimate, and so on (cf. Locke 1973). Developments in modern liberalism have, however, shed doubt on this postulate. When, among others, John Rawls argued that the distribution of natural endowments and handicaps among humans was arbitrary and could result in an arbitrary distribution of goods within society, a distribution that required compensation of some sort, the 'self' was in fact cut off from its endowments, making the latter at least in theory as much part of nature as trees, soil and ore. Now whether this abstract 'self' or anyone or anything else can acquire and possess anything at all – the self's endowments as much as any other part of nature's stock – is open to question. Rawls's own theory is ambiguous at this point. He assumes on the one hand that no one deserves the capacities with which we are born, so that these natural endowments can be seen as resources for society – and society then decides on the development and use of its citizens' endowments, plus whatever they gain through their use. Michael Sandel has correctly pointed out that this is a *non sequitur*: just because I do not deserve my Nobel prize qualities as a writer does not mean that society does (Sandel 1982: 77). We need a further argument to allow society, or in fact anyone, to take possession of faculties and goods, use them and distribute them. But on this topic, Rawls is silent. He seems to implicitly accept John Locke's view that mixing our labour with nature can make a product our own – where in this case 'we' is society rather than an individual, since society ultimately 'owns' the means by which nature is transformed into goods. However, he gives us no reason to believe that mixing our labour is a sufficient condition for the creation of ownership rights. What is needed here is an argument to justify original acquisition – looting nature – *as such*, regardless of whether society or the individual performs the act of acquisition. In classical liberalism and in some libertarian circles (cf. Nozick 1974), the solution to this problem is to think of original acquisition as a natural right. Once we reject the natural rights assumption (as arbitrarily as Nozick accepts it),[18] we are left without any justification of original acquisition, of any subsequent transfer of goods and of the use we make of them (i.e. the precise permissions associated with 'ownership'). Since I have no intention of getting entangled in a debate for which I see no resolution, I shall ignore this problem here and join the *communis opinio* among liberals in simply assuming that there can be such a thing as legitimate ownership of our natural endowments. At the very least, we must admit that one individual will have at least as much right to the ownership of their specific endowments as anyone else. Moreover, it should be kept in mind that attributing the self with ownership rights to its endowments is in itself an empty move: it tells us nothing about the permissions associated with this. As a matter of fact, and at least in so far as rights to goods other than endowments are concerned, we do not seem to need any postulate about *self*-ownership rights to justify other ownership rights.

It is far easier to defend the idea that *individuals* can legitimately have property. To paraphrase William Galston: even though neither the individual nor society deserved an individual's natural endowments, the way a person directs the development of their talents should be considered a private achievement (Galston 1991: 131). Society or my social environment influence me but they do not make me; they merely create the favourable circumstances in which, or the duress under which, I develop and employ my talents. Hence, and other things being equal, society or my causers in society cannot claim credit for the use I make of my talents – only I can. Note that this establishes a *principal* justification of individual ownership rights only. It does not exclude that at least those others who contributed to my achievements can lay claim to that part of my achievements for the employment of which they are responsible, provided that they were not already morally obliged to do so. Nor does it exclude that there may be *other* reasons to attribute rights and goods to individuals or collectives.

We may then assume that there can be individual ownership rights. We have as yet no reason to believe that there can be ownership rights to everything, but there is reason to believe that *some* things can be owned. Finally, there is as yet no reason to believe that an ownership right to x implies that we can do everything we like with x, i.e. that ownership is an unconditional right, one with which no one may interfere, or that all rights are by definition inalienable.

I can be relatively brief about the latter point. Morally assessing the effect of a transfer is not fundamentally different from assessing that of original acquisition. There are very few physical objections to the transfer of rights: I can give you an apple almost as easily as my labour, my thoughts or helping hand. Now if I can take a piece of nature and call it my own, I can also give it back, so that you can take it and call it yours. A transfer from me to you is the same, the only difference being that we do not make the detour through nature. The question in both cases (original acquisition and transfer) is why a reasonable impartial person should (dis)approve of the conditions under which X (you) can come to own P (my apple). There seems to be nothing wrong with me selling you an apple, but suppose that I own the Amazon rain forest and voluntarily sell it to a building company that plans to turn it into one gigantic Milton Keynes, or that I sell my help's two hours of labour to a neighbour, or that I sell you the right to use my body for vivisection. In each of these cases we can doubt whether the new owners are the *morally legitimate* owners of my forest, my help's work and my body. The important thing, however, is that these questions relate to the 'who owns what' question, to the issue of which rights to P I have and which society – but they do not call the possibility of alienability and legitimate transfer into question.

Individuals can thus have ownership rights to goods and these rights need not be inalienable. The next question is: can these rights be absolute – can we do anything with them we like? In practice, a world of merely absolute rights would be impossible to live in: anyone or any set of individuals who can save lives but refuses to would legitimately make life hell. There is a case for believing that at times an ownership right will be a right to use rather than (in this sense: absolutely) possess a thing, and to use it only in particular ways, places, periods or circumstances. What makes many rights conditional is the fact that no matter how scarce or abundant they may be, as long as they are alienable, claims to them can be compared, evaluated and ordered. And even in the absence of competing claims a claim need not become a right. The fact that only one person applies for, say, a subsidy is not enough reason to give it; we also need a positive reason to recognize their claim as valid. Hence the strength of arguments determines whether we should agree to a particular attribution of rights. Because we have no yardstick for the ultimate truth or validity of principles (see Chapter 1), we have to accept, first, that any right that can physically be withdrawn can also be withdrawn morally should a stronger argument come along, and second, that a right to x is not a right to x-plus – a right to pluck apples is not a right to cut down trees. In other words: ownership of a tree does not include a right to chop it down or take the apples *unless* either right was explicitly granted.

In sum: rights can be alienable and if they are, they need not be absolute. We should therefore distinguish three hypothetical spheres of rights: rights that are both absolute and inalienable; unconditional rights (absolute but not inalienable); and conditional rights (neither absolute nor inalienable). Given our criteria of liberal democracy, what rights will we find in each of these spheres?

Let us start with a rough sketch of the sphere of inalienable (and absolute) rights – some important details will be added later. For liberalism, respect for individuals and their plans of life is the cornerstone of a just society. For the self, it is impossible to execute a plan of life without any endowments. Taking away a person's *natural* endowments seems to be physically possible but only within very strict margins: there is, for instance, a limit to the number of organs we can spare and there are some we really cannot live without. Hence, there is a class of natural endowments that cannot be taken away for plain physical reasons without ending the self's existence and with it any perspective on fulfilling its plan of life. The same conclusion must be drawn if taking away some natural endowment results in the loss of a necessary condition for subjecthood (agency or consciousness). Apart from these two types of endowments, there seem to be no goods that cannot be alienated in principle. Unconditional rights would, in a liberal theory, be associated with goods that can be alienated yet are essential to the realization of a plan of life. Without wanting to go in to too much detail, examples could be rights to nourishment and to having a say in politics, since both determine how much room is given to one's plan of life. Unfortunately, the objects of unconditional rights can be (made) scarce – there is, for instance, only a limited amount of food available at any given time. Even if food is essential for x's plan of life and x has a legitimate claim to food, it does not follow that food should be provided (cf. Barry 1995a: 72).

Finally, conditional rights would simply be all rights to all other goods.

Our rough-and-ready categorization would be far too rough without the addition of at least six notes. Plans of life differ. What may be an essential natural endowment for one (say, kidneys for a drinker) or an essential alienable good (say, vodka for the same) need not be essential for another person. Hence, we cannot describe inalienable, unconditional and conditional rights in too much detail. Moreover, plans of life change over time: there may come a day when vodka is no longer one of my basic needs in life and a healthy liver is. We may therefore have to allow some leeway in categorizing rights.

Plans of life are not necessarily monolithic. They can have a core of relatively permanent basic goals or side-constraints, criteria for a full life, next to a less important series of auxiliary goals that determine how the full life should be lived under specific circumstances; the latter need not give rise to unconditional rights. If your overall goal in life is to be rich (core), you should invest in gold today (periphery); if your aim is to die as a sincere person in everything you say and do (core), you should not enter a career as actor (periphery). If individuals themselves do not already make this distinction between core and periphery, then it exists at least in the minds of political theorists as one between basic needs and secondary wants (cf. e.g. Barry 1995a: 72 ff.). Furthermore, I have categorized goods, not the precise acts that we may use them for. We may implicitly assume that an unconditional right to nutrition does not entail a right to toy with and throw food because the latter right does not at first sight seem to be essential to anyone's plan of life – but it could be, for some eccentric at least. Conversely, our categorization allows kidney transplants since humans can live without half of them, yet it does not make transplants mandatory: an obligation to let a body-snatching government transplant a kidney from a healthy person *a* to a sick *b* could well violate another of *a*'s rights, their right to self-determination, which might be one of their basic needs. Again, although our categorization allows us to attribute different types of rights in different cases to different person, it cannot be listed comprehensively.

The categorization given here is incomplete in one other important sense. It has been modelled to fit human beings, not subjects in general. If applied to animals, the idea of a plan of life loses its relevancy even though the categories can still be applied to their case (*mutatis mutandis*). Consequently, animals can still have inalienable and unconditional rights to the goods (room, food, wings, legs) that make them conscious and/or agents at the moment they are conscious agents but – exceptional circumstances excluded – no right to the continuation of their existence because they have no plans for tomorrow, i.e. no more interest in exist-ing tomorrow than in not existing. Finally, it is worth noting that very few goods fall into the categories of inalienable and unconditional rights, and that those that do are necessarily quite abstract. A right to nutrition does not necessarily imply a right to that one particular cabbage at that precise moment in time. Most if not nearly all goods around us, nature for starters, are the object of conditional rights. If we want to know how a liberal democratic society should treat nature, it is at this sphere that we have to take a look in the first place.

5.2 The restraint principle

This section has as its topic, in formal terms, the attribution of rights to condi-tional goods or, less formally, the distribution of goods to satisfy wants. Initially, I hold on to the assumption made in the previous section that it is legitimate to distinguish between (basic) needs and (further) wants, the first determining a subject's unconditional rights and the second their conditional rights. Before defending this assumption, I want to introduce a particular kind of principle for the attribution of conditional rights: a side-constraint on distribution called the restraint principle.

Attributive principles can be of two kinds: one determines what goods or rights are to be distributed, the other how they are to be distributed. The first puts side-constraints to distribution, the second defines allocatory aims. The latter are the kind of principles to which liberal theorists refer as 'principles of justice' in a strict sense. Both can offer ways towards an environmentally sane society. On the one hand, green interests *can* perhaps be satisfied better by one scheme for the alloca-tion of resources than another, and liberalism *may* support such a scheme, but there is no reason to assume beforehand that either thesis is true (I shall discuss them in Chapter 7). On the other hand, there can be side-constraints to the distribution of rights to natural resources that ensure, say, sustainability regardless of the way in which resources will be allocated, so that we do not have to worry too much about the environmental effects of liberal principles of justice.

It is this latter type of principle that I shall discuss here. I argue that liberals should accept a principle that demands that conditional rights to (in a physical sense) scarce goods be distributed in such a way that they remain, within the limits of necessity, available for redistribution. This restraint principle limits the set of rights to a good x that are available for allocation by excluding, in principle, the right to destroy x. Thus, for example, a liberal democratic society can allow 'private ownership' (or collective ownership; the difference is immaterial) of a work of art, say *Guernica*, in the form of a series of rights to hang the picture wherever the owner wants it, to look at it or not, to cover it or not, to sell it or not – but it cannot give the owner rights to burn it or cut it up. Now replace Picasso's painting by Sherwood Forest and we end up with a principle that protects nature against arbitrary and unnecessary destruction. So much for the claims, now for the arguments.

The restraint principle demands that we do not distribute any rights to the destruction of objects of conditional rights, in other words:

> no goods shall be destroyed unless unavoidable and unless they are replaced by perfectly identical goods; if that is physically impossible, they should be replaced by equivalent goods resembling the original as closely as possible; and if that is also impossible, a proper compensation should be provided.

The hypothesis that has to be proved here is that there are things we just should not destroy unless using them is unavoidable and there is no other way to use them but by destroying them. Even then, we ought to try to replace them by the best possible substitutes.

Now the attribution of rights is a matter of arguments. We need *reasons* to acknowledge one another's rights to goods, otherwise a distribution scheme would be arbitrary, accidental and (since it would be as good as any other) extremely temporary – in the absence of good reasons in favour of one distribution, there is no reason why it could not be replaced at any moment by any other. Claims to rights can be compared, evaluated and ordered – and even in the absence of competing claims a claim need not become a right. It is the strength of arguments that determines whether we should agree to a particular distribution of conditional rights or to a principle of distribution. Because (as liberals) we have no yardstick for the ultimate truth or validity of principles, we have to accept, first, that any conditional right (any good) that can physically be withdrawn (alienated) can also be withdrawn morally should a stronger argument come along, and second, that a right to x is not a right to x-plus – a right to pluck apples is not a right to cut down trees.

The things we call rights in everyday life are, after all, very complex sets of what we have called molecular rights, i.e. permissions, duties and prohibitions relating to 'basic acts', to use one specific object to one specific purpose in one specific way at one moment in time and one place in the universe (cf. Section 1.3). The rights involved in owning, say, a forest then consist of a long range of 'molecular' permissions describing what we may and may not, should and should not do at any time with each of the elements of which the forest is made up. As in real life, the owner can, for instance, be free to hunt rabbits in it but not foxes, free to let it grow naturally or cut the dead wood, free to sell it but not to fell it. In short, they can be allowed to use it but not in every way they might want to.

The recognition of (formal) rights to each basic act requires, in principle, a separate argument. It is simply not true that an argument p that would convince us that I should have the freedom to type or not type 'e' next will convince us that I therefore also deserve the freedom to vote. An argument p may support rights to more than one basic act, but p does not support my right to do x_2 merely *because* it supports my right to do x_1. By the same token, and other things being equal, a valid argument for picking fruit from trees here and now legitimizes only those acts that are necessary to the purpose.[19] It is not necessary to cut down the tree to get hold of the fruit, therefore a right to possess or take possession of the fruit does not entitle us to cut down the tree. Arguments in favour of a specific set of complex rights can only bring us so far and no further.

If an ownership right (in the colloquial sense) to x does not imply that we can do everything we like with x, it becomes possible that there are specific molecular rights to x that simply cannot be granted, as a matter of principle. As a matter of fact, a case can be made for the thesis that there is at least one such right that can never be granted: the right to destroy the object of x, thus turning ownership rights into rights to use rather than (absolutely) possess x.

Conditional ownership depends on reasons, but reasons are as fallible as those who articulate them. The soundness of reasons may change over time or depending on the information we have. We may believe that asbestos is safe, thus grant the right to use it in houses; that phosphates are safe, and grant the right to use them in washing powder; that nuclear energy is safe and allow the construction of a nuclear power plant near Harrisburgh. In due course, experience and research may lead to different estimates, thus forcing us to withdraw our earlier decisions – as long as and in so far as they have not actually been carried out.

Others may turn out to have better claims to goods than I have – they merely have not yet made those claims, have not been able to do so, or have not been heard. They may even be unaware of their own good reasons, as it takes time to grow up and develop or discover our plan of life. In this context then, a first-come or finders-keepers principle is not a warrant for justice: my claim may be prior in time without being prior in terms of urgency or need. In sum: whenever there is a choice between destroying a good, thus depriving others of present or future options to realize legitimate plans, or merely using it without limiting other people's options, we have a duty to chose the latter.[20] Consequently, we are 'custodians rather than owners of the planet' (Barry 1977: 284).

In theory, then, no argument can legitimize that we destroy anything that could be used by anyone else for a better reason. Theory is often an unpractical thing. Following a general duty not to destroy anything means starving or thirsting or freezing to death – whichever comes first. No environmentalist who values the lives of subjects (humans or other) and no liberal who value the liberty of life more than anything else could ever accept this. Hence, both will agree that this duty to restraint can be overruled by necessity, i.e. by the quest for survival or, in liberal terms, by the quest for a life worth living.

Although necessity overrules the duty to restraint, it does not annul the rights and claims of others whose options we limit by turning x, say, Yellowstone Park, into farm land. In general, your valid claim to x does not make it a duty for us to make an alternative for x available to you, yet it does not prohibit it either: we are, in principle, free to replace Yellowstone Park but not obliged to. However, there are claims and claims. If x is vital to my survival or to my liberty of life, it can also be vital for you. There may be reasons why my survival is less important than yours – which reasons exactly is not important here. Depending on the specific ethic of specific theorists, they could be that 'you' are a whole people and I am a mouse, that I love you and we both value your life more than mine or that I will die soon and you still have a long life ahead of you. Anyway, your claim to x can be stronger than mine, only your claim has not been recognized for the same reasons that we saw could legitimize the duty to restraint. In circumstances like these, the freedom to replace x turns into a duty to replace – with the same proviso: unless necessity prohibits it.

Both in theory and in practice, then, the best way of saving the spirit of the duty to restraint is to demand that an object should not be destroyed unless unavoidable, that if it must be destroyed it should be replaced by an identical object, that if this is impossible an equivalent object should be made available, and that if the last is also impossible, a proper compensation should be provided. In this way, we always offer the best possible substitute and leave as much room as possible to protect the interests of other claimants – whose claims, after all, may be better.

There is still a hitch in all this.[21] The demand that nothing be destroyed 'unless necessary' might form a weak spot in the restraint principle. In point of fact, it depends on the contents of the specific liberal theory of justice where exactly the border lies between necessity and indulgence. As far as the restraint principle is concerned, the problem is not where this border lies but whether it makes sense to draw a border like this at all.

In everyday life, we actually do make a distinction between what I would like to call (basic) needs and (further) wants. Extreme situations excluded, we would say that 'mere survival' is a pure need and 'the fiftieth Rolls Royce' a pure want. What justifies this distinction – both in everyday life and in theory – is, for one part, the fact that some but not all goods, acts, rights, etc. are necessary conditions for a tolerable life and necessary but not sufficient preconditions for a morally good, rewarding or full life (cf. Barry's (1995a: 11, 84) conception of crucial goods). What the latter consists of is, again, up to the individual theorist to decide. Even the terms 'need' and 'want' are not sacred: a similar distinction can be made in terms like 'use' and 'abuse' or (perhaps more to the green taste) 'symbiotic' and 'parasytic' (cf. Barry forthcoming). We also have a moral duty to care about the availability of these necessary preconditions more than we have for the presence of less important goods. Only if these basic conditions have been met can humans be free to take responsibility for their own behaviour and actions – only then can they be moral agents. Assuming, then, that the difficulty of defining a precise criterion of 'need' or 'necessity' cannot count against the in itself sensible idea of a distinction between basic needs and further wants, the restraint principle can be as valid and sensible as any other.

It is worth while considering some of the implications of this conclusion. A first implication should not be too surprising: if there are rights that can only be distributed on a conditional basis, then property rights can no longer be considered sacrosanct. We already know that they are not absolute in real life and positive law. We now also have confirmation that they need not be absolute from a moral point of view: that is, that they cannot only be distributed but also redistributed or even withheld.

Moreover, the restraint principle, as a moral principle, points to some serious shortcomings in real life with regard to the preconditions present society creates for future generations. (After all, the green critique of life in modern society does not accuse it of destroying nature but of destroying too much of it.) The fact that the restraint principle allows us to criticize practice from an 'internal' liberal-democratic point of view rather than from an external 'green' perspective is, in this context, perhaps even less important than the discovery that these stains on the reputation of liberalism are for real.

On a more modest scale, note that the restraint principle creates a link between liberalism and the debate in green political theory on the preconditions for a sustainable society. Note, for example, that the restraint principle protects more than humans only: it also incorporates non-human nature. Let us assume that animals are not subjects but mere objects, livestock. As argued in Section 4.3, the notion of instrumental value is compatible with strong, weak, absurdly weak and even outrageously weak substitutability – in other words, even a merely instrumentally valuable object can be irreplaceable. There can hardly be more protection for 'natural objects' like animals. The restraint principle demands that, unless there is no way to avoid it, no rock, animal or plant should be destroyed, no species made extinct. It also demands that if this is impossible, the most similar possible alternative should be made available: nature should be replaced by nature, not by concrete. It puts the onus of proof for the legitimacy of environmentally harmful acts on the bad guys. Finally, and this brings me to the subject of the next section, the restraint principle offers a quite elegant solution to a problem discussed in both liberal and green theory: that of justice between generations, particularly between existing and therefore morally relevant generations and non-existing and therefore not necessarily relevant generations. It protects the interests of future generations by simply protecting those of present generations.

5.3 The savings principle and the restraint principle

There is more to the restraint principle than meets the eye. It can be defended formally, as I did in the last section, as a strict implication of the idea that rights must rest on arguments. It can, however, be defended in other ways as well. One possibility (see Dobson forthcoming) is to build on John Locke's defence of private property: if original acquisition is justified only if (among other conditions) 'enough and as good' is left for others and whatever is taken from nature is used for needs, subsistence or the greater good of the community rather than to stock and rot, then there seems to be ground to assume that restraint in exploiting nature would at least be prudent. However, I want to explore another line of argument here.

One of John Rawls's most original and fundamental contributions to the theory of social justice has been the savings principle (Rawls 1972: 128). In just four sentences and a footnote somewhere at two-thirds in his latest book, *Political Liberalism*, Rawls replies to his numerous critics of the preceding 20 years with an elegant new defence of this principle (Rawls 1993: 274). I hope to show (1) that the new defence is better than the old one, although still not unproblematical; (2) that it can be generalized, i.e. that it is also relevant to liberal theorists who do not want to work within a Rawlsian framework of original position and contract theory; (3) that the acceptance of a Rawlsian savings principle is a necessary condition for a just liberal society; and (4) that the savings principle logically implies the restraint principle.

When he first introduced the idea of a savings principle, Rawls rejected the notion that individuals in the original position have obligations to their immediate descendants simply because they do (Rawls 1972: 128). A move like that might make persons in the original position far too distinctly individual and would certainly contradict the basic assumption that people in the original position are mutually disinterested. He did, oddly enough (cf. Thero 1995), assume that (abstract) next generations have valid claims towards (abstract) present generations. This assumption justified the introduction of a further motivational assumption: people in the original position were to be imagined as representatives of bloodlines, 'as being so to speak deputies for a kind of everlasting moral agent or institution' (Rawls 1972: 128). On the other hand, these representatives (to whom Rawls refers on other occasions as heads of families) are biased in a way: they are expected to represent the interests of at least the next two generations but not their bloodline's 'entire life span in perpetuity' (Rawls 1972: 128). As a rather straightforward implication of these assumptions, persons in the original position will agree that a savings principle should be included in the list of principles of justice. Yet Rawls finds it impossible to specify this principle beyond the point of saying that it depends on a society's level and rate of development (Rawls 1972: 287) – partly because there is little sensible to be said about the preferences of future generations (Rawls 1990).

Although his defence of the savings principle has been severely criticized for countless reasons, Rawls has answered only one point directly, i.e. that there would be a contradiction within the set of assumptions that make up the original position. Apart from the fact that people in the original position are not supposed to know how old they are or to which generation they belong, Rawls assumes both (a) that everyone in the original position is mutually disinterested and that (b) everyone cares for their descendants' fate. Condition (a) implies that the contracting parties will not be interested in any kind of savings principle at all, condition (b) implies that they will be. This appears to be a contradiction, and Rawls has admitted as much (Rawls 1993: 274n.). In reply, Rawls made two very simple moves: he drops condition (b) and redraws the outlines of the original position in such a way that the contracting parties will 'agree to a savings principle subject to the further condition that they must want all previous generations to have followed it' (Rawls 1993: 274). I shall refer to this new condition as condition (c).

Condition (a) turns out to do nearly all the work. People in the original position may be disinterested in the fate of others, but they care a lot about themselves. It is precisely to this feeling that the new original position is expected to appeal: it makes the savings principle mutually beneficial. Let me give a first and deliberately imperfect illustration of this: if your grandparent offered you £5000 on the condition that you promise to give your grandchild £5000 at say the same age, and the alternative is to start in life with nothing but the clothes in which you were born, it would be irrational not to accept your grandfather's offer.

Handing down Rawls's interpretation of pounds, primary goods, from generation to generation rather than money from parents to children is far more ingenious: it implies promising to hand down goods to a next generation, *regardless* of whether we have or want or like children. It is, moreover, not just a fair weather principle: Rawls demands that the savings principle be formulated in such a way that it would be accepted by all generations, even if economic times are rough. The savings principle will have to be flexible enough to guarantee that no generation will be worse off *relative* to any previous generation: that is, relative to the circumstances in which each generation lives. The new defence is furthermore foolproof against breaking promises made to dead people because promises are made to generations rather than individuals. Then again, it is not foolproof against a generation that decides to take the money and run, unless we assume that these people have a sense of justice or obligation. Fortunately, Rawls already assumed the existence of 'a sense of justice' in his old original position (Rawls 1972: 567).

At this point, two questions naturally arise: 'What's to stop, precisely, the generation in question from taking the money and running?';[22] and 'on what sort of psychology is that generation-skipping concern systematically more plausible than the original story about caring about our immediate descendants in Rawls I?'[23] The answer to both is simple: rationality.

Rawls makes one important improvement on the orthodox understanding of justice between generations[24] – visible, for instance, in condition (c): generations exist next to each other, not one after the other: 'society is a system of cooperation between generations over time' (Rawls 1993: 274) – it definitely does not say 'a system of transfers from one generation to the next'. Although this more realistic conception of generations does not change the substantial role that trust plays in social cooperation, there is no altruism in it, no other-regarding attitude: just self-interest and mutual disinterest. Since generations exist as contemporaries, defecting generations impose a cost on society of which they will have to pay a part: defection destroys the basis of trust on which society was built. The other generations can perhaps go on as before in their cooperative venture but they will have learned that no one can be trusted – a lesson that cannot be too healthy for the survival of society.

To put the point more systematically, Rawls assumes in condition (c) that new generations enter an already ongoing system of social cooperation. On entry, they have a choice between accepting the existing rules (the principles of justice chosen behind the veil of ignorance) or rejecting them. Rejection would mean losing the social bond and with it the advantages of social cooperation. Since it would be irrational to reject a system of rules that does not make us worse off and might actually improve our situation, the new generation will accept the existing rules, including the savings principle. Once included in the system of cooperation, any generation can choose at any time between keeping its promise or breaking it. They will only do the latter if they stand to gain from defection. Being mutually disinterested *individuals*, they will set off the losses incurred by leaving society against the gains of either individual exclusion and solitude or a life in the company of other untrustworthy defectors. Under the standard assumption that the primary social goods are rights and liberties, opportunities and powers, income, wealth and self-respect (Rawls 1972: 92, 441), it would be irrational for any individual to step out of society and lose these goods. Moreover, being reasonable persons in the original position, they would devise a system in which they would not want to risk losing out if others – or even a whole generation – were to step out of the contract. Hence, the contract rests on general mutual consent and not on one-sided promises. In all these considerations, the elementary factors are rationality, mutual disinterest and trust in the other's rationality. Altruism has no role in it.

Note that the possible existence of as yet non-existent future generations is irrelevant to the Rawlsian argument. There is no way to protect them directly against 'thieves' – stealing from them is easier than stealing candy from a baby. Instead, they are protected indirectly at any given moment by a principle that means to protect the interests of existing generations. Incidentally, this saves Rawls from having to discuss the objection (cf. Narveson 1978: 58) that we have obligations to the next generation we create because we put them on the spot, but not to any third generation our descendants create because that is *their* responsibility.

There is one serious problem, though. We cannot assume that a well-ordered society already exists or that people in the original position design principles for a society that already meets the right principles. With his condition (c), Rawls explicitly assumes that the system of cooperation is already in existence: the savings principle will not be accepted unless we can want previous generations to have accepted it. However, it seems that a first generation would only lose by accepting the savings principle, hence it would not be motivated to accept the arrangement. The original position has no first generation. The veil of ignorance precludes knowledge of the generation to which we belong, so that given Rawls's standard assumptions about the original position, people in it will nonetheless accept a savings principle. Yet there are first generations in real life. Every change in the structure of a society, every move towards a more just society, is made by a first generation that will (ideally) judge its own performance in relation to what people in the original position would decide. If the latter cannot represent a situation which the former experience on a regular basis, a choice situation that will influence their well-being, then there may well be something wrong with the reflective equilibrium between our considered convictions and the conditions of the original position.[25]

The (I think) only answer that avoids an appeal to altruism would run as follows: since all next generations will 'want' the previous generation to have agreed to the savings principle, their mutual basis of trust would fail to come about if one generation defected – in which case all generations, including the first one, would lose out. And given Rawls's list of primary social goods, it seems reasonable to assume that the costs of defection will far outweigh the benefits, even for a first generation. To put it bluntly, it is in granddad's interest to invest in his grand-child; he thereby gains the trust and support of two generations.

Rawls has this time given us an internally perfectly consistent justification for the choice of incorporating a savings principle in the just society (cf. Barry 1977: 280). The sheer mutual advantage of such a principle ensures that mutually disinterested persons behind a veil of ignorance will choose it; the combination of rationality and mutual advantage ensures that it will be a stable principle.[26] But Rawls has done far more. First, given the assumptions on which the original position is built, he has implicitly shown that a savings principle, hence a form of justice between generations, is a necessary condition of the just society. Since the savings principle operates to the advantage of every next generation, it would be irrational not to choose it. Conversely, its absence from a society would make that society unjust, i.e. biased in favour of previous generations. Such a society would lack a basis for trust and hence for its stability and survival.

Second, by representing generations as contemporaries, persons in the original position know that for most of their lives, their well-being will depend on generations that are not even around yet. Thus, by ensuring justice between existing generations, Rawls has implicitly also ensured justice towards non-existent future generations. Perhaps this goes two steps further than Rawls himself wished to go, but it is an obvious and unavoidable consequence of his train of thought.

Third, he has given us an argument that can be generalized to fit a far wider range of liberal theories of justice. Justice between generations, in the form of a savings principle, turns out to be an essential trait of nearly every liberal theory of social justice. I shall prove this by first weakening the conditions of Rawls's original position, thus extending the argument to liberal contract theories in general. As the idea of a contract is a vehicle for moral arguments rather than an argument in its own right, the conclusions reached can subsequently be shown to be valid for all liberal theories of justice with only one exception.

First, then, we will have to revise all the conditions Rawls needs in order to support a savings principle – that is, conditions (a) and (c) as well as others – and replace them by more general conditions:

Table 5.1 Premises for liberal savings principles

Rawls needs	Liberals need
1 Mutually disinterested individuals who,	1 Rational individuals (i.e. goal-maximizers) who,
2 in order to fulfil their reasonable plans of life	2 given a plan of life that satisfies some condition of admissibility (reasonableness),
3 prefer more rather than fewer primary social goods, which are	3 prefer more rather than fewer of those rights that are defined as relevant by their plan of life; rights to goods that are
4 relatively scarce, and	4 relatively scarce and
5 can be made available on a wider scale through social cooperation.	5 become available through social cooperation.
These individuals are endowed with:	These individuals are endowed with:
6 the knowledge that there are (simultaneously) similarly conditioned previous and next generations,	6 the knowledge that they are contemporaries of similarly situated next and future generations,
7 a sense of justice,	7 a common basis of trust: the willingness to justify at least a possible defection from the scheme of social cooperation,
8 a veil of ignorance that covers their identities,	8 either some, or all, or none of the elements of the veil of ignorance,
9 general scientific knowledge about the structure and nature of societies,	9 all available relevant information, which they use in a rational way,
10 a preference for the maximin principle,	10 the legitimate authority to establish, (re)distribute and limit property rights, assuming
11 the legitimate authority to establish, (re)distribute and limit property rights.	(a) that there are distributable conditional rights, and
	(b) that arguments for the just distribution of specific rights may be overruled by better arguments.

Provided the first set of conditions apply, persons in Rawls's original position will want a savings principle to be established and they will want it to take such a form that 'they must want all *previous* generations to have followed it' (Rawls 1993: 274; his italics). (Note that 1 to 6 imply Rawls's condition (c).) Provided the second set applies, individuals in any liberal contract theory will want the same.

Most of Rawls's conditions can be rephrased without any difficulty in terms that fit other liberal theories of justice. Condition (2), for instance, is innocent enough to fit any conception of liberalism since they all understand the individuals' plans of life, in some form or other, as the core value of liberalism. We can simply leave it up to individual theorists to define 'reasonable'. This first change in the structure of the original position being made, the logic of Rawls's argument still stands: what matters is not the content of plans of life or whether they are admissible in the original position, but the fact that there is something relevant at stake for persons choosing principles of justice.

Conditions (3–5) can be adapted in a similar way. All we need is a common expression for the means required to fulfil plans of life, a concept of which 'primary social goods' is one conception among many. For this purpose I shall use the term 'rights'. Let us assume that social justice is about the distribution of unspecified things (the specification depends on the theory and theorist) that can be distributed in the form of rights. In other words, we distribute goods and represent them as rights, but we could have chosen another term – say, claim, good, benefit and burden, vehicle of utility, etc. We shall also assume that rights are distributed to individuals only. The idea that individuals are the primary recipients of justice fits in best with liberalism, since liberals usually only and at most care for collectives as preconditions of the individual's freedom to pursue their plan of life (cf. Kymlicka 1989, 1995).

Since individuals are assumed to care for their plans of life, they obviously also care about having the rights necessary to that purpose, and about having more rather than less of them. Furthermore, the assumption of relative scarcity (where 'relative' can be further specified within a particular theory) is standard – there is little practical use in talking about distributive justice under conditions of abundance or extreme scarcity.[27] Assumption (5) is also standard within liberal theories of justice and needs no further revision. Again, these changes do not influence the basic structure of the argument in favour of a savings principle.

We can move beyond the requirement of general scientific knowledge (9) by demanding that persons in revised original positions have access to all the relevant information available and use it in a rational way – again leaving it up to the specific theory to determine whether 'rational use' means 'complete and comprehensive use' or whether there is a limit to the amount of information that contracting parties can cope with. Again, this new assumption is incorporated in some form or other in all liberal theories of justice and does not change the structure of Rawls's argument.

It seems that we cannot change the requirement of a veil of ignorance (8) without also adapting condition (1), mutual disinterest. The more parts of the veil are dropped, the more mutually disinterested individuals become rational individuals, goal-maximizers. They thus lose their impartiality and start to design principles to their own immediate advantage, and that does not obviously include an interest in the fate of future generations. However, condition (6) warrants that persons in an original position will know that a stable system of social cooperation safeguarding their liberties and opportunities can only succeed if, at the same time, it satisfies enough other parties who may be members of other generations. The latter may not be a common assumption in liberal theories of justice (they often do not even discuss the topic of intergenerational justice) but it is both intrinsically reasonable (because realistic) and in no way incompatible with liberal views of justice. Hence, even without (all elements of) the veil of ignorance, it will be reasonable to accept *a* savings principle: everyone (except the first generation) stands to gain from it. Note, though, that the way in which the contracting parties are characterized in specific theories will certainly influence the *form* of the savings principle. In a society of perfect altruists, for instance, the appropriate savings principle might well require perpetual starvation of every generation for the sake of every next generation. In a society of perfect egoists, on the other hand, next generations would receive nothing more than their bargaining power allows them to extract from their predecessors.

Note that the extension of conditions (1) and (8) to rational individuals with or without elements of a veil of ignorance has one disturbing consequence: it allows for coalitions and majoritarian decision-making on principles of justice. Societies can, after all, usually survive the disappearance (or in this case, non-inclusion) of one or more of its members. Nonetheless, even a society of minimal size contains (by assumption) more than one generation; hence, the structure of the argument still remains intact. As for the sense of justice (7) which gives Rawls his basis for trust, we do not have to demand more than that the contracting parties will not break agreements without good reason; being rational, the defector knows that defection brings less gain than costs.

The one but final condition that can be extended concerns the contracting parties' preference for the maximin principle (8). We shall adapt it by dropping it. In fact, the maximin principle is redundant in Rawls's argument for the *idea* of a savings principle but it deserved mentioning because it affects the precise *form* it can take. It has been argued, for instance, that the Rawlsian version of a savings principle would copy the difference principle (in the defence of which the maximin rule plays a crucial role): persons in the original position would, or should if Rawls had done a better job, distribute goods over generations to the benefit of the least advantaged, i.e. the powerless non-existing next generations (e.g. Wells 1995). This is not the place to discuss this idea – let us just note that the defence of the idea of a savings principle does not depend on a specific attitude towards risks, but on the pure rationality of gaining anyway (relative to a society without savings principle).

The last condition to be adapted is (11), which requires that contracting parties have a legitimate authority to establish, (re)distribute and limit property rights. It is only here that we meet a real difficulty. Not all liberals, especially not libertarians, if liberalism is understood in a broad sense, will accept that anyone other than the possessing individual has the right to use or abuse their property. Given our discussion of (un)conditional rights in the previous two sections, there can be only one answer: rights to goods require arguments and an appeal to natural rights is insufficient, no argument except an appeal to natural rights can make every right unconditional, therefore conditional rights can exist. Thus, condition (11) can remain unchanged provided two further assumptions are made: we have to assume that arguments for the just distribution of specific rights may not be final, an assumption that should be built into distributive principles; and we must assume that the set of distributable conditional rights is non-empty. The (untenable) type of libertarian theory in which property rights are assumed to be absolute do not fit into the picture; they cannot contain a savings principle except one based on the voluntary compliance of individuals.

We have moved from Rawls's own contract theory to Rawls's theoretical clones, first and second cousins and beyond, and each time when we extended the conditions of Rawls's original position, we saw that the validity of the argument for a savings principle remained unaffected. By now, the argument has been shown to work for virtually every liberal theory that can be represented as a contract theory – the exception being libertarian theories in which ownership is a natural right, and we have rejected the latter as unsound. It is now possible to drop the contractarian façade and explain why *any* liberal theory of justice, contractarian or not, should contain a savings principle. What the ten conditions above describe is a typically liberal view of society: goal-maximizers with a plan of life, for the realization of which they need private property rights, rational individuals who stand to gain from social cooperation. In the presence of persons from different generations, it is simply rational for goal-maximizing individuals to seek the mutual benefit of a savings principle. This makes justice between generations, and the savings principle in particular, a necessary condition for the stability and (therefore) survival of a liberal society.

It is important not to lose sight of the limitations of this account. We had to assume that nature may legitimately be exploited and that exploitation can give rise to legitimate ownership rights. We still know nothing about the precise contents of the savings principle. More importantly, we cannot extend the validity of the argument beyond liberalism. If, for instance, individuals and their plans of life were products of society or incarnations of a culture (as they are in communitarian theories), the appeal to mutual advantage, self-interest and individual liberties and opportunities loses its force.

The implications of these few sentences on page 274 of *Political Liberalism* are nevertheless quite interesting. We now have reason to believe that Rawls can defend his savings principle, that liberals in general need to include a savings principle in their respective theories of justice – and that (some form of) obligations to future generations is a *conditio sine qua non* of any liberal theory of justice. The obvious question then is: what form will this obligation take?

One answer is that it would have to take the form of a substantive principle, one that describes exactly how much of which goods we owe to future generations. Green political theorists have suggested several such principles based on the politically popular demand for sustainability: that is, the demand that we 'meet the needs of the present without compromising the ability of future generations to meet their own needs'.[28] In all these, the aim is to ensure that humans ought to pass their planet on to future generations 'in at least no worse shape than they found it in' (Barry 1977: 284; cf. Routley & Routley 1982: 123). I shall refer to this condition as 'no worse off' (NWO). So as not to complicate matters too much, I shall ignore the so-called identity problem: different policies will create future generations made up of different people whose well-being can only be compared to non-existence (Barry 1977: 281). I shall furthermore assume that we are only saving material goods for future generations, although the argument can be extended to cover immaterial goods – institutions – as well. Unfortunately, all this is not enough to make these principles viable. We will, I argue, have to opt for a more formal principle – like the restraint principle.

Let us start with the uncomplicated idea that the savings principle would be satisfied if present generations leave the world no worse off than when they entered it. As it turns out, NWO might well be acceptable for people in a Rawlsian original position but not under most other circumstances. It is a fair weather principle, assuming the possibility of the continuous growth of welfare or at the very least a steady state economy. In deviating cases, and regardless of what other arguments can be offered in its support, NWO is not an optimal answer – not to our question, and not to the demands of green political theorists for a principle that ensures sustainability. Imagine the following oversimplified situation: we have at t_1 a generation g_1 with a welfare of w_1. Instead of welfare, we can read freedom, primary goods, sets of rights, options. Our g_1 that has the opportunity of investing in the future in such a way that it:

(a) may improve but will not lower its own welfare: $w_1 \geq w_1'$
(b) leaves the subsequent generation g_2 worse off than it would have been without investments, i.e. $w_2 < w_2'$, and
(c) leaves next generations $g_3 \ldots g_n$ better off than they would have been: $w_3 > w_3'$, $w_4 > w_4', \ldots w_n > w_n'$ (cf. Beckerman 1994: 200, 1996).

As an illustration, imagine that the first generation cuts down all its man-made forests, sells the wood and leaves the land in a derelict state. The next generation will have to live with wasteland, the generations after that will have access to both economically useful wood and aesthetically attractive wilderness. (Contrary to Barry's opinion then (Barry 1977: 277), we can make our remote successors better of by making our immediate successors worse off.) From the point of view of a green theorist this situation is to be preferred to the one in which NWO had been applied. The same happens to be true for the goal-maximizing individuals described in the previous section. As a rule, they cannot lose but stand to gain from violating the principle. NWO makes them worse off any time an investment disadvantages one single future generation or perhaps even one single future individual. Thus, they would never choose this as their savings principle. Even Rawlsian risk-avoiders would have to reject NWO: putting themselves in the position of g_2, they would have to accept NWO since it leaves them better off, but NWO would lead to subsequent generations being worse off than they could be, so they would have to reject it as well. Rawls's minimax condition and NWO are logically incompatible. Note, finally, that NWO takes no account of the population size. Imagine that g_2 comes into the world finding the same amount of resources its predecessors found, only the population size has doubled. It will have to live in misery if it is to hand over exactly as much as it found to the third generation.

Let us now adapt NWO and turn it into a pareto-optimal principle: generations should only invest in future generations if that leaves no one worse off and at least one person at least as well off as would otherwise be the case. The first thing to note is that this principle demands, at least, that persons in an original position are mutually disinterested in each others' fate. Egoists have no reason to accept pareto-NWO. In addition, nothing substantial changes as regards our example: pareto-NWO will leave generations g_3 and following worse off.

One step forward would be to accept that generations can legitimately be made worse off, under conditions. Consider, third, this pareto-minimum rule: no generation should leave the subsequent generation in a position in which it cannot lead a life that is better than sheer misery. The implication of this rule is that each generation should see to it that no next generation falls below a certain basic level of welfare and that, apart from this, anything goes. Consequently, every generation is free to rob its successors of endless other opportunities they could have had – which will be unacceptable to anyone who risks ending up in a later generation.

Finally, consider a combination of these three rules: generations should not be made worse off than preceding generations, unless the overall gains outweigh the overall losses, and provided no generation falls below a minimum standard. Apart from the principle's insensitivity to population size, and apart from the obvious problem that we might have to work with an infinite number of future generations to establish whether there is an overall profit at all, this rule will certainly not be acceptable to risk-avoiders. Moreover and generally speaking, it allows us to sacrifice the well-being of one generation to a perhaps minute gain for others in cases where all generations could perhaps be very well off.

The restraint principle, on the other hand, is something completely different. It can only be overruled by the quest for survival, i.e. if there is no other way to preserve a present life worth living other than by harming a future life. As a consequence, the restraint principle does exactly the same work as we intended NWO to do – it leaves no one worse off than physically possible – without the disadvantages attached to NWO. What is more, it is a logical implication of the argument for a savings principle: it is consistent with individual self-interest, even to the degree that it minimizes risks for risk-avoiders, and it ensures that future generations inherit as much as they can without present generations losing anything essential. The restraint principle can be as valid, as sensible and as necessary for the survival of liberal democracy as the principle of which it is an interpretation: the savings principle for justice between generations.

6

Population policies

6.1 Strategies of sustainability

One of the least contestable remarks about sustainability is that it is an essentially contested concept (Jacobs 1995). Among the many interpretations or, in Rawls's terms (Rawls 1972; cf. Gray 1980), conceptions of it (for more see Dobson forthcoming), the following at least must by now sound familiar: weak, strong, absurdly strong and outrageously strong sustainability. A second series of sustainabilities – ecological, economic and social sustainability – describes various criteria for sustainability. The meaning of these conceptions is fairly obvious, yet transforming them into measurable criteria may be quite difficult. Ecological sustainability is perhaps the most limited conception: for an ecology to be sustainable, the presence of humans is not required. Economic sustainability on the other hand does presume the existence of humans (without people no economy); it also presumes at least a minimal form of ecological sustainability in that it requires the continued presence of those natural resources that cannot be substituted by others. Finally, social sustainability (cf. Dobson's sustainable society (Dobson 1995: 17)) seems to be the broadest conception of the three. It not only demands economic sustainability and the minimal form of ecological sustainability included in the latter, but also a sustainable culture and structure of society, which may pose further limits to humankind's exploitation (in the neutral sense of the word) of the natural environment. Despite the differences, all these conceptions of sustainability share a series of 'core ideas'. To those already mentioned (particularly obligations to future generations) I should now like to add two series of core ideas about the source and the parameters of the problem of sustainability. Rather than copying worn analogies like Spaceship Earth or Lifeboat Earth, I want to present them in the form of a new metaphor, the *mescal tub thesis*. Mescal, for those who did not yet know, is a Mexican liquor distilled from cacti; a special kind of hairy worm is added to the liquid to (unfortunately) absorb the hallucinogenic substance mescaline which the cacti contain:

(1) The world is like a tub: a (nearly) closed system containing a limited amount of substances.

(2) Humankind is one of the worms living in this tub, which is filled with mescal; we drink it, greatly enjoy what it gives us (mescaline) and secrete what we do not need. And we sometimes eat the other worms (animals, plants) for the mescaline they contain.

(3) Since we live in a tub, our capacity for 'producing' mescaline is limited by the available resources; there are definite limits to growth.

(4) Incidentally, the ratio secreted/raw mescal (waste/resources) is visibly increasing.

(5) The limits to our growth dictate that we cannot sustain ourselves unless we use our resources sensibly, unless we eat and drink less and more efficiently, recycle our waste, or slim down.

The mescal tub thesis, or the set of ideas for which it stands, also contains the second set of core ideas, the *parameters* of the sustainability issue (cf. De-Shalit 1996:10; compare Nelissen, Van der Straaten & Klinkers 1997: 318; Harrison 1997):

(6) The supply side parameter(s) 'resources', usually split up into two (natural, artificial), three (artificial, irreplaceable and replaceable natural) or more distinct parameters. This aspect of sustainability will be discussed in Chapter 8.

(7) The demand side parameter(s) 'consumers', as a rule only the next few generations of humans, but the needs and interests of animals or biotopes (etc.) are sometimes included as well. Consumers are the topic of this chapter.

(8) The parameter 'distributive structure', describing the relations of access to and use of all types of resources by all types of consumers. We shall discuss matters of distribution in Chapter 7.

If we look on the world in this way and also recognize the interests of 'consumers' or subjects as insuperably or uniquely relevant, it is only natural to seek solutions on the basis of the concepts of this framework rather than to reject or redesign them, especially since the framework appears to generate sensible answers. As long as it does, as long as the answers can be implemented in the real world and as long as there is hope that they will work, there seems to be no reason to consider alternatives. In fact, unchallenged belief in the mescal tub metaphor or, other guises of virtually the same theses, in Lifeboat Earth or Spaceship Earth makes alternatives like Starbase Earth (in which we would all board spaceships and roam the universe for new planets to consume) look silly. This is both true for mainstream politicians and environmentalists who interpret sustainability in a very limited anthropocentric sense as well as for deeper greens who shun the term sustainability but nevertheless accept the thesis that there are limits to growth.

Yet we should keep in mind the fact that the position taken on one green issue does not necessarily commit us to any particular position on others (cf. Chapter 2). Hence, on the one hand, sustainability is a 'minimal' green concept: any course of action (solution) it prohibits will certainly be prohibited by deeper green theories that accept the existence of limits to growth, and among the actions it allows will be those acceptable to deeper greens, although not all these permissible actions will be good enough for them. On the other hand, we can be green without belief in limits to growth: understanding environmental problems as the purely moral question of how humanity should relate to nature presumes that there are limits to what should rather than what can be done to nature. On the latter perspective, the parameters of sustainability and the conclusions drawn from a limits to growth perspective are simply irrelevant. Whereas the preceding chapters were open to both interpretations, this and the following two chapters are written from a limits to growth perspective only. For the 'moral' green who does not believe in empirical limits to growth, nearly everything that can be said about the possibility of greening liberalism has already been said. With the exception of the effects of the restraint principle on questions of distributive justice (Chapter 7), nothing can or needs to be added. For 'limits to growth' greens, on the other hand, these three chapters either add ideas that must be relevant if they are also 'moral' greens, respectively contain the only ideas that can be relevant if they understand the environmental crisis are a purely empirical problem.

Given the three parameters just described, the solution strategies are obvious. We can distinguish three approaches which may occur in various combinations. Sustainability theorists who focus on the distributive structure will look for fair distribution schemes that maximally satisfy legitimate demands. Any solution, any proposed set of distributive principles Q and criteria for the legitimacy of demands P, will be intended to be general in two senses: it is to be applicable to any morally permissible permutation of the set of possible conceptions of 'resources' (see core idea 6 above) and to any corresponding permutation of 'consumers' (core idea 7). On a second, more technocratic approach, 'consumers' and 'distributive structure' will be treated as given and the question of sustainability will be dealt with as a matter of developing alternative resources, discovering new ones, enlarging efficiency, promoting recycling. On the third approach, typical of among others 'deep' green thinkers, lifestyles and life plans can be criticized, as can the genesis of lifestyles itself (culture and education) or simply the volume of consumers' demands.

In this chapter, the first of three devoted to the three dimensions of sustainability, I shall discuss demand-side solutions and concentrate almost fully on population policies. In Section 6.2, I examine several ways to influence the parameter 'consumers' and conclude that only population size, i.e. the number of consuming and (re)producing individuals, appears to be a feasible policy subject. I then affirm that there are good reasons to evaluate population policies not only from the macro (quasi-)welfare perspective but also from the micro (individual) rights perspective.

Using the formal conception of rights developed in Section 1.3, I translate the notion of procreative rights into these terms in Section 6.3; among other things, I shall distinguish four distinct rights involved in procreation. I argue here that procreative rights should at least in principle be individualized, i.e. that they ought to be attributed to the potentially procreating individuals themselves. Nevertheless, there are also good reasons why procreative rights should not be absolute (read: completely individualized). There is, therefore, even on a rights perspective a theoretical possibility to defend population policies.

Section 6.4 applies our series of stringent criteria for liberal democracy from Chapter 2, for each of the rights distinguished earlier, to a series of reasons for limiting them in the cause of sustainability. I shall inspect the case of legally enforced antinatalist measures rather than that of less controversial indirect policies (taxation, propaganda), and this for four reasons. Apart from the fact that arguments against the former are usually voiced in extremely vague and general terms as 'obviously' incompatible with a Western view of individual freedom (Bayles 1976a; Walter 1988), the more precise type of argument that I shall develop appears to apply to both types of policy. Moreover, if direct antinatalist policies cannot ensure a sustainable population level, we can be sure that indirect policies, since they depend on control over a broader range of variables, will fare at least as badly. Finally, the case of strong antinatalist policies offers an excellent opportunity to expose the sometimes perverse logic of the antiprocreationists. It is in this section that I introduce the 'rapid reproducers paradox'. I shall argue that antinatalist policies will invariably punish and reward the wrong people.

Having discussed the compatibility of *stringent* conditions of liberal democracy with population policies, and having concluded that the former offer little room for the latter, I end in Section 6.5 with a discussion of three antinatalist solution strategies, beginning with indirect policies (incentives, deterrents, fines, etc.). The second strategy consists in relaxing the conditions of liberal democracy, i.e. in compromising liberty, equality or democracy for the sake of survival. The third involves the employment of policies other than those directed at population size. Without wanting to give away too much of the argument, my overall conclusion will be that, even on a rights perspective, we cannot preclude the possibility that the quest for sustainability may force us to act wrongly in the pursuit of a presumably good goal.

6.2 The idea of a sustainable population

One way of dealing with the sustainability problem thus seems to be that of influencing the parameter 'consumers', and one way of doing this is by limiting the volume of the consumers' demands. Now the question of volume is not irrelevant to other approaches, since there will nearly always be a lower and an upper limit to the volume of satisfiable demands. There is obviously a mathematical upper limit even in circumstances of nearly infinite availability of resources, infinitely small pollution and the most stringent distributive conditions P and Q; as long as resources are not abundantly available and environmental costs absent, there will be a level at which the mere volume of admissible demands outweighs any supply. Conversely, there is at least in terms of social and economic sustainability a lower limit, a minimum number of people required to keep an economy or a society viable. (As my editor remarked, there are indeed endangered peoples, as there are endangered species.)

There are a number of ways in which volume control might be thought possible – even though the means may be repulsive. The actual size of consumers could be modified. As a matter of fact, it has been argued that the increasing length and, partly in consequence, weight of the average European human since the Second World War has been responsible for a 25 per cent increase in food consumption, merely because larger bodies need more food (Samaras 1995). Obviously, large people also need more cloth, more warmth, more energy, taller houses, more of nearly everything, all extra and apparently quite legitimate strains on a limited stock. If the assumption is correct that the cause of all this is the gigantic rise in welfare since the beginning of this century, i.e. higher incomes, healthier lives, better food, better medical care, then it would appear that decreasing growth might have an environmentally fascinating side-effect in terms of the size of individuals.

We could also change people's diets, for instance by banning meat, as has been argued by vegetarians and radical animal rights activists (Gold 1995). The production of consumable proteins, vitamins, etc. via the meat of animals appears to be quite inefficient in environmental terms. This strategy could be applied more radically by not just excluding the wrong kinds of food but prescribing the right taste or, one step further, by diminishing people's appetite.

Alternatively, in line with a perverse anti-growth policy we could develop an equally repulsive death policy to counteract the environmentally harmful consequences of the incredible rise in human life expectance – from 30-something five centuries ago to nearly 60 (in the West nearly 80) today.

Finally, there is always the possibility of influencing the number of consumers, i.e. the size of the population, which happens to be our subject. Applying one or both of the first two techniques, even if feasible, will not by itself provide any final solution to the volume control problem: we can imagine nearly infinitely small people with the smallest possible appetite for the environmentally most efficient greens only, and nevertheless discover that there can be too many of them to sustain. Hence, there still is reason to ask if the size of a sustainable population can and should be a subject of environmental policy. (I assume closed borders; the issue of immigration and emigration will be discussed in Chapter 8.)

Now what exactly a sustainable population size is depends on our definition of sustainability. There is reason to believe that the most adequate understanding of sustainability in our context is that of social sustainability. I think we can safely assume that whatever reasons we have for wanting our natural environment to be sustainable will be reasons that include the positive value of the continued existence of humankind: the reasons that would make the Chinese tiger or the Byzantine unicorn worth preserving will only on the most radical interpretations of ecologism exclude humans. Hence, it takes more than just a sustainable environment if humankind is to survive. It also takes a sustainable economy to produce the goods that will enable humankind to survive and reproduce and, under the assumption that a life must be worth living if the idea of sustainability is to make sense, it takes a (reproducing and developing) society, at least to safeguard its economic survival and quite probably also to provide the kinds of welfare an economic system cannot or need not provide.

As asserted above, there are two ways of addressing the population size problem: from a familiar quasi-welfarist macro-perspective and from the less familiar (individual) rights perspective. I refer to the former as *quasi*-welfarist because it is an amalgam of two strands of thought, one typically non-green and utilitarian (the Parfitian treatment of population problems) and the other typically green and positivistic. The latter is welfarist only in its prescriptive role which is secondary to its main, descriptive interest in the carrying capacity of ecosystems. The most conspicuous common denominator of these two views is that their 'unit of analysis' is a complete population. The rights perspective, on the other hand, is interested only in individuals. It should not be confused with a *natural* rights perspective. The conception of rights that will be developed here can, but need not, represent natural rights. It is a relatively neutral analytical instrument that can be used to represent natural rights equally well as individual interests or individual welfare.

My reasons for focusing on the rights perspective are twofold. First, even if we suppose that the macro-perspective is uniquely valid or that it overrides any obligations arising from the other (rights) perspective, a quasi-welfarist answer to the right *size* of the population may still leave the question of the *distribution* of procreative activities unanswered; there may be valid reasons other than overall welfare why some people would be more deserving than others.

There is also a more positive reason to examine population policies from a rights perspective. I do not believe that the quasi-welfarist perspective embodies the only or most important moral code. Individual rights are important too, and they may pose limits to either the means or the goals of a (welfarist or quasi-welfarist) population policy. The reason is based on, I believe shared, intuitions, and therefore not a final proof – if there can be anything of the sort in moral philosophy – but it does seem to give us a prima facie reason to recognize the legitimacy of this individualistic perspective.

Now if individual rights matter, they must have some impact on the morality of policy decisions, either as limits and side-constraints or as goals; that much is clear. The deeper question is why individual rights would be so important, and the answer to this has little to do with the notion of *rights* since rights are merely a formal mode of representation of that which really matters about individuals. Next, why exactly individuals matter is open to debate. We need not agree with the choice I made earlier to justify moral relevancy in terms of plans of life, agency and consciousness. The history of philosophy has given us many alternative answers ranging from human attributes like natural rights, happiness and telos to reason, free will, interests and even welfare. The point is *that* there is something about individuals which matters. Even if individuals are sacrificed in a greater cause, we believe in that cause because of what it means to, ultimately, other *individuals*. It is individuals who are either the vehicles of whatever is good or the good in itself, and it is precisely because we believe something like this that we find a principle like *fiat justitia pereat mundus* perverse. Please note, by the way, that the individuals whose fate matters need not be identifiable presently existing humans only. The argument refers to individual entities as such, regardless of the qualities that distinguish the morally most relevant ones from whatever other entities exist.

The right to procreate or, as Garret Hardin so compassionately called it, the freedom to breed (Hardin 1976) seems to be one of the most sensitive subjects of debate in society. Let me just give one illustration. During the first months of 1996, members of the Enviroethics list on Internet were involved in a long and interesting but after a while slightly tedious debate on the meaning and relevance of the concept of intrinsic value. On 8 March, one member changed the agenda by voicing – rather loudly – her fury over the imminent extinction of the (Chinese) tiger, indirectly accusing her audience of being engrossed in useless academic twaddle. In the next few days, several other often violently angry messages and reactions were sent out. In answer to the suggestion that saving (human) lives might sometimes justify the killing even of the last members of a species, some argued that the survival of a whole species was far more important than that of individual members of, to quote one, 'a superabundant species'. In ensuing reactions, one person stated that she would defend her child's life against whatever threat, even the last tiger, by any means necessary; a second remarked that it was exactly the uncontrolled urge of humans to procreate that had brought us (he probably included the tiger) to where we were now. At the moment the right to procreation was questioned and the ferocity of the reactions reached unprecedented heights, the moderator of the list was forced to interrupt the debate and call the contributors to order.

Procreative rights may be a sensitive and delicate subject, yet that is no reason not to discuss the subject 'academically', in all senses of the word. Procreative rights are – the Enviroethics debate illustrates this point as well – undoubtedly relevant to several problems of sustainability. They are directly related to the issue of the relative moral importance of present and future generations and hence to that of the just distribution of resources between generations. Since procreation relates to the attribution and exercise of (individual) rights, including a prima facie right to the use of resources, it is also in itself a matter of distributive justice and sustainability. Most obviously, the distribution of procreative rights influences the size of a population (at least in theory) and thereby the actual size of the problem of sustainability.

Procreative rights are a slightly underrepresented subject in the literature on population size and population control. There is, admittedly, a large body of literature written from a utilitarian perspective. Much of it deals with Parfitian people problems, which are derived from Parfit's famous discussion of whether causing someone to exist can be good in itself or only circumstantially, and, if the latter, then relative to which measure or alternative (Parfit 1984; Attfield 1988). Yet as a rule these texts deal neither with rights (but rather with welfare) nor with sustainability (though they could be applied to the subject). There is also an amazing number of publications combining philosophical or theoretical considerations with empirical data on population size and sustainability (e.g. Hardin 1976; Walter 1988; Irvine 1991) – though not necessarily all Doomsday writings (compare e.g. Graham & Graham 1992; Tabah 1995). Yet with sustainability or alternative concepts operating as the supreme value and the macro level as domain, the issue of rights is either again largely ignored or straightforwardly dismissed as being overruled by considerations of survival. What remains is little more than a dozen or so articles and books over the last, say, 30 years. Given the deep and possibly extreme intrusions on people's lives, sex lives and bodies that would be involved in population policies, this is another reason to attempt to draw attention to the moral status of procreative rights.

6.3 Procreative rights

Above, I made the claim that the right to procreate is not so much *a* right but rather a whole class or set of rights; now I am going to sustain that claim. To begin with, procreation itself is not a single basic act. Procreation is a (chrono)logically ordered set of basic acts, each of which can but need not be performed (read: CanDop, *and* CanDo¬p,), and each of which is, in accordance with our criterion for a basic act, 'only contingently related to the rest' (Steiner 1983: 76): that is, there is more than one alternative intentional cause of the same effect. To list only those basic acts involved in 'classic' procreation (with apologies to the sensitive reader):

(1) We are free to use or not use our sexual parts for 'normal' intercourse. No necessary prior act is required, and alternative forms of intercourse are imaginable.
(2) Women are free to be or not be fertilized by means of 'normal' intercourse. There is nowadays an alternative called artificial insemination. Note that there is no such distinct basic act for men; they have only one means of fertilizing.
(3) Women are free to carry or not carry a child; an alternative to fertilization as prior cause is implantation of an already fertilized egg from another source.
(4) Women are free to carry a child through to birth or not; an alternative is abortion.

Note that I do assume all sorts of fortuitous empirical circumstances to obtain to the effect that we *can* actually physically perform these separate basic acts. The inclusion of amoral, non-legal obstacles to procreation (impotence, infertility, the time between ovulations) would, however, only complicate our analysis (we would have to consider rights to the impossible) and add little more than that to the results. Furthermore, I disregard all possible complications that may arise out of the fact that since a newborn child is a person (or vehicle of value) it may also have been a person (or vehicle of value) at least some time before its birth. Finally, I take the liberty of assuming that postnatal persons themselves cannot be (should not be) the subject of a 'modest proposal' policy (infanticide) aimed at population control.

A population policy can be aimed at either one or more or all of these four basic acts, and it can put limits on the time, place, aim, means and intention for the enjoyment of rights to them, as well as restrict their enjoyment to certain individuals only. Obviously, some policies will be more effective than others – a temporarily continuous ban on basic act number one, for example, will (if obeyed) have at least as much effect as forbidding one individual to enjoy the same on 1 April 1999 between 4 and 4.10 a.m. on the top of Mount Everest. However, apart from allowing us to assess the possible efficacy of any given population policy, an analysis in terms of formal rights can do more. It can help us point out when and where and for which reasons a particular assignment of procreative rights can be justified.

Before we can discuss the legitimacy of particular restrictions on procreation in the cause of population control, we should first determine if there can, as a matter of principle, be restrictions on these rights at all. We can answer this positively, even without an appeal to some higher principle that would override individual rights. Speaking intuitively and in terms of legal rather than formal rights, it seems quite reasonable to prohibit rape, lesser offences like the unauthorized use of sperm from a sperm bank and dangerous behaviour like having sex while driving a car – as reasonable as prohibiting torture, theft and drunk driving. In more general terms, the exercise of one legal right may violate other rights or another person's rights, and we do not need a standard for the relative importance or value of distinct rights to conclude that p and $\neg p$ cannot be the case at the same time and place, that one or the other right has to give way, i.e. that it is quite sensible to put restrictions on the exercise of rights. Finally, if we translate this idea into the language of individual formal rights, some technical changes must be made: in this vocabulary, rights cannot 'conflict', first because p and $\neg p$ cannot be the case at the same time and place, and second because formal rights are not first given and afterwards limited, but are instead already restricted from the beginning to molecular rights, i.e. to one individual, one time, one place. In these terms then, the argument would be that because two individuals cannot (by definition of an individual formal right) have the precise same right, individual rights cannot be assigned as absolute rights: that is, they cannot be assigned to all individuals, for all time, any place, to any means, or for any intention or purpose. At least not in principle.

If we are to discuss the fair or proper distribution of procreative rights, we also need a moral standard. To this purpose, I shall use a limited form of our series of stringent criteria for liberal democracy. Many conditions, like those of free association, freedom of thought and freedom of contract, are only marginally relevant to the issue of procreative rights; I shall ignore them here. Those which at first sight seem to be most relevant are (and note that 2–4 are side-constraints on 1):

(1) Decisions on procreative rights should express the ideal citizens' will: they should be sincere, based on consideration of all available information and of possible consequences, prudent, responsive to needs and directed at a common good.
(2) The numerical political equality of citizens should be respected.
(3) The proportional equality of recipients should be respected, i.e.:
  (a) equal cases should be treated equally and unequal cases in proportion to their inequality; and
  (b) there should be a reasonable measure for equality: that is, one that is supported by good reasons, open to debate and open to amendment should better reasons emerge.
(4) The 'liberty of life' should be respected: individuals should be free to design and pursue any reasonable plan of life and lifestyle.

6.4 The attribution of procreative rights

Before we turn to the issue of assigning procreative rights, some remarks should be made. First, note once more that we do not 'distribute' formal rights but assign them; in other words, we want to recognize morally valid claims and reject invalid ones. Second, we shall, as usual, assign rather than *limit* rights, at least in principle. The formal representation of rights used here is built on the notion of (permissions regarding) single basic acts; only compounds of such permissions, complex rights, can be absolute or limited. Moreover, the expression 'to limit a right' seems to presume the prior existence of a (more extensive) right, and it can be used in legal discourse, but it would be out of place when assigning or recognizing rights. It would be contradictory to assume the existence of, say, legal rights prior to their codification. The type of question we have to answer then is not 'Why limit Natasha's rights?' but 'Why assign Natasha a specific right to procreate?' The answer to that will have to be phrased in terms that address the following conditions:

(1) Natasha must belong to a class of subjects who are able to procreate;
(2a) assigning Natasha a specific type of right to procreate is inevitable given our moral standard; or
(2b) this type of right reflects our moral standard better than any other type, and there is no other individual for whom that particular type of right to procreate would better reflect our moral standard;
(3) Natasha's specific type of right to procreate p is compatible with any other individual's specific type of procreative right r, the latter being at least as legitimate a right as Natasha's right to p.

We cannot, at least not in principle, deal with this issue by asking if there are good reasons why we should limit Natasha's procreative rights to particular times, places, means, end or intentions.[29] In principle, we should ask if there are good

146

reasons to assign each of the types of right to each particular basic act, or if there are better reasons in favour than against. However, in practice, we do not need to go so far as to discuss every individual's single basic acts. There is a good reason to assume that each and every individual capable of procreating should have a universal, absolute and complete liberty to do so: anywhere, any time, for any reason. That good reason is our moral standard itself, according to which, first, all individuals should be treated as equals unless good reasons for unequal treatment exist. Hence, the onus of proof for the legitimacy of inequality rests on those arguing against equal treatment. Second, respect for the liberty of life demands that individuals should have the liberty to perform each of the four basic acts involved in procreation, since – an important assumption – each can be part of a reasonable plan of life, anywhere, any time. Observe that this is a rather strong assumption; we shall further analyse it in the next section.

Hence, we can, after all, ask if there are good reasons to limit any individual *i*'s procreative rights. We already have a criterion for establishing whether a reason is a *good* reason – our moral standard. What we need next are reasons themselves, to be specific: reasons relating to the population question. Fortunately, 30-odd years of debate have given us more than enough of those – very weak reasons, as we shall see. In the next few paragraphs, I introduce and discuss a list that pretty well reflects the conventional objections to unlimited procreation. Please note that the authors I mention as sources for the arguments are merely sources and not necessarily advocates.

1. The interests of the procreating individual(s) themselves are not served by procreating

Supporting arguments for this claim can come from several sources. We might argue that, according to a cost-benefit analysis, purely personal circumstances like financial position, health, etc. would worsen if the procreating individual(s) took a (further) child. We might also point to the possibly doubtful ethical status of their instrumental reasons for taking a child: using another person (the child) for reaping the benefits of caring for, raising and educating the child. However, even if these could count as valid arguments against procreating, which we seriously doubt, they are not specifically *environmental* arguments. From that perspective, only the personal experience of environmental degradation, the loss of resources or natural capital, can count as an objection.

One of the interesting features of this last argument is that if it were valid, it could only be used to limit some but not all procreative rights; another interesting feature is that it happens not to be valid. As to the first point: we really have to stretch our imagination to believe that the extra use of resources and hence the possible environmental degradation due to the first three basic acts, intercourse, fertilization and pregnancy, could ever be more than negligible. Hence, our first argument can apparently only be used to turn the liberty to carry a child through to birth into a duty to abort. It would require additional arguments against (state-enforced) abortion to put – less obviously cynical – limits to other liberties instead. Fortunately, the effect of procreation on the procreators' interests is not a good argument; it cannot pass the test of our moral standard. People's plans of life are their own affair and they are not open to political debate. It is up to the prospective parents to decide if they want to carry the personal environmental costs of a (further) child, and it is even up to them to decide if they want to care about personal costs at all.

2. The interests of one's own new child are not served by its creation
(Narveson 1976; Attfield 1988)

The supporting argument for this is essentially the same as that for the first: environmental degradation, reduction of resources or of natural capital. Consequently, the same restriction applies: if valid, the argument can only serve to justify a duty to abort. A first and obvious problem with this argument is that it seems to presume the impossible: a comparison between the quality of the new child's life and the quality of its life if it did not exist. There seems to be no point of reference for a sensible comparison, as Parfit once pointed out (Parfit 1984). Parfit also provided an alternative reading of the possible person problem, according to which we could argue, in our context and given a constant size of natural capital, that the addition of one person j to a population of N might lead to such a decrease of the averagely available portion of natural capital that the quality of j's life would lie below a level at which j would consider life worth living. At this point, we encounter a second problem: j will be incapable, for at least some time after their birth, to judge on the current and expected quality of their life. But since the state, the polity or the legislator cannot judge on this one person's (possible and future) plans of life, let alone on the quality of the resulting life, without – at least in principle – depriving everyone else of the liberty of life as well, there can be no legal duty to prevent j's birth.

3. The interests of (other) existing persons of the same generation or age group as the procreating individuals are not served by creating a new individual
(Attfield 1988)

4. The interests of the next (already existing) generation or age group are not served by adding one more child to their number
(Attfield 1988)

Since the supporting argument and restriction of these two claims are again essentially the same, we can immediately move on to the objections. As far as the third argument is concerned, we shall initially assume that all members of this generation are capable of procreation.

The fact that these two arguments refer to the interests of *others* raises the question whether one individual from an age group can be allowed to procreate and another not, given that they have equally worthy plans of life and presumably share at least some interest in a sustainable environment and an adequate level of resources. Unequal treatment of these cases would seem to violate the criterion of equality. The one difference between the two cases that might be relevant is time: if individual *i* was born before *j*, and if we may expect *i*'s lifestyle to leave 'enough and as good' for others, whereas the addition of *j* would be one step too far, then perhaps *j*'s existence should be prevented. Notice, first, that this crucial distinction between sustainable *i*'s and 'superabundant' *j*'s is intrinsic to the logic of any theory that presumes the existence of a finite quantity of resources, mescal tubs, Spaceships Earth and Lifeboats Earth included: at any given moment in time, there must be a limit to the number of people who can be fed and kept alive. Notice also that the distinction between *i*'s and *j*'s is descriptive, not prescriptive. Being the first to finish, i.e. being the creator of *i* rather than *j*, is just not enough to justify the moral or legal right to be the only one to finish.

If we accepted the birth of *i* and the undesirability of the birth of any further *j* as facts of life, life not always being fair, then a decision to prevent the existence of new *j*s and so add to the unfairness of life a portion of voluntary unfairness, creates a new problem: that of compensation. Prohibiting individuals to exercise rights that would, in other circumstances, be considered perfectly legitimate to exercise calls for compensation. But it is, first, not clear who should be compensated since it is not clear either who should be prevented from procreating. The threat *i*'s creators pose to the welfare of *j* and *j*'s creators is just as real and deep as that of *j*'s would-be parents: either one or either couple prevents the other from putting a child in a viable environment. And the same is true for the last child before *i*, and the one before that. Since *i*'s creators could just as well have been *j*'s and vice versa, blaming one couple rather than the other is still arbitrary. Moreover, even if prohibition could be justified and compensation required, we may seriously question if there is any way to compensate a parent for 'losing' a child. If there is any form of natural capital that seems utterly insubstitutable, it must be a human individual.

Any attempt to formulate a fair way of prohibiting procreation for the reasons mentioned would demand equal treatment of all concerned – say, a general ban on procreation 'for as long as necessary', which would seem to harm everyone equally. But even this does not solve the problem. There is, actually, a limited sense in which a temporal universal ban on procreation can be described as fair: it gives the members of the existing next generation mentioned in the last argument a reasonable basis to build a life on and it puts no member of that generation at a disadvantage. But there are drawbacks. The ban limits the liberty of life of one present generation in favour of the next, which may still be intergenerationally fair if we believe that the first generation has already had the fullest opportunity to lead a rewarding life and that the next generation now deserves the same chance – if only that did not presume that whatever rewards life has in store for

us are either measured out before procreation or distributed evenly over life. Moreover, even if our ban were intergenerationally fair, it would nevertheless still be
based on intragenerational unfairness. In particular, it still harms those who would
like to have (more) children more than those who already have (enough), and it
is still biased in favour of those who accidentally (or deliberately – who knows?)
finish first. There is also a practical problem: announcing the introduction of a
population policy may be counterproductive: it may actually incite 'rapid reproducers' (the parents of the *i*s) to grab their last chance at reproducing, rather than
deterring anyone.

Finally, consider the interests of the members of the procreating generation in
the third argument who choose not to procreate or simply cannot. In their age
group, they are the only ones who cannot in any way be blamed for causing the
existence of the *j*s that are too much and the *i*s who make the *j*s too much. They
have better reasons to complain about the creators of the *i*s and *j*s for limiting
their life prospects, and perhaps also for having given precedence to non-existing
persons (the new generation they created) over existing people. Yet even if this
complaint were made and justified, it would at best offer grounds to compensate
the childless, but it is insufficient reason to support any limitations to the right to
procreate. The latter would also require a good reason to believe that the interests of at least the involuntarily childless should count for more than those of the
procreating individuals and their offspring. Moreover, it would bring us back to
where we were before: at the inadmissibility of all solutions to the rapid reproducers problem.

(Note that, under really extremely special circumstances, the rapid reproducers
problem does not arise: if we could give everyone an equal right to reproduce at
a rate of *n* children per person, *n* being constant, a natural number and the rate at
which a population has to renew in order to remain a maximum-size sustainable
population. This also requires other factors to be constant: the size and appetite of
individuals, the level of resources, technology and so forth.)

5. The interests of future generations or persons are not served by procreation (Callahan 1976; Norton 1982b; Attfield 1988)

Let us assume, for the sake of clarity, that this argument would be used against
the prospective creators of *j*s (the 'superabundant' children) and in defence of the
interests of the future offspring of the *i*s. One aspect in which it differs from the
fourth argument is that it concerns people who do not yet exist. In the ordinary
moral analysis of environmental problems, which is focused on group, species and
societal effects, future generations can offer an extra reason for environmentally
sane behaviour. Here, in an analysis of the individual aspects of a possible population policy, it turns out to add little more than complications.

The problem with future persons here is, for once, not so much that they and
their interests do not yet exist but rather that they do not *need* to exist. This is
true both for the superabundant *j*s and the third generation created by the *i*s.

150

What we have here, then, apart from the already discussed conflicts among existing generations, is a conflict of interest between two non-existent groups. The first question we should answer now is whether we should be impartial between these two generations, partial to those in the nearest future or partial to generations in the more remote future. This question is relatively easy to answer. There is no relevant difference whatsoever between the *j*s and the children of the *i*s: either group is equally contingent, equally 'willed' into existence, either group is equally non-existent now, and we are equally uncertain about the needs and interests of either group. We must, therefore, be temporally impartial in this matter.

It follows that we cannot hold any *one* generation hostage to the threat of a miserable fate for any future generation, any more than any *other*. The *j*s of today are equally morally relevant as, and equally pernicious to the welfare of, tomorrow's children of the *i*s, and the latter generation will in turn face the same problem once it reaches the point where it could create its own *i*s and *j*s. (And so on into the far future; cf. Norton's 'Distance Problem' (Norton 1982b).) In short, the fifth argument leads to a continuous repetition of the arguments made against the third and the fourth; it provides no grounds for a limitation of the individual liberty to procreate.

6. *The interests of other existing species are not served by procreation*
(Attfield 1988)

7. *The interests of future generations of other existing species are not*
served by procreation (Norton 1982a; Attfield 1988)

8. *The interests of (the rest of) nature as it exists are not served by procreation*
(Attfield 1988)

9. *The interests of (the rest of) nature in the future are not served by*
procreation (Attfield 1988)

At this stage, the antinatalist is thrown back on rather dubious arguments against procreative liberties. For these four arguments to be valid, we have to believe, first, that animals or non-human nature can and do matter as morally relevant subjects, and second that they are at least as relevant as human individuals – either as individual creatures, or as sets, species, parts or in sum. Let us for once simply grant the former condition and concentrate on the latter.

In the first place, it seems obvious that we cannot make sense of the eighth and ninth arguments without reference to the concept of intrinsic value. If we were to use the notion of instrumental value, we would have to presuppose that cauliflowers have something equivalent to interests, which might stretch the common understanding of 'interest' slightly too far, not to mention that accepting such an exaggerated view of interests would shed doubt on the morality of eating greens. I shall discuss the alternative reading of these arguments in terms of intrinsic value below as the argument that follows.

Hence, only the interests of animals might result in viable interpretations of our second condition: to limit human procreative liberties, either individual animals or sets (say, families or herds) or a species as such should be morally more valuable than individual humans. Since we are still discussing instrumental value, we can also exclude sets and species; whatever value they have, on this account, is instrumental to the welfare or interests of individual animals. Next, remark that if we want to defend a ban on human rather than animal procreation, animals should in fact not just have the same moral significance as individual humans but, at least occasionally, be rated higher – a claim which I doubt non-utilitarians (cf. Attfield 1988) would want to defend and which, if Bryan Norton is right (Norton 1982a), utilitarians should not want to defend.

Nevertheless, assume that animals or non-human nature matter enough to support the claim that at least some humans should not procreate, at least not for some time. Whatever else may be true, it still does not follow that humans (or animals!) should be *legally* limited in their procreative rights – and this for the very same reason why the interests of other humans are insufficient to override these rights: any limitations of the right to procreate will result in an unjust distribution of the burdens of sustainability, profitable only to those who procreate fastest.

10. The intrinsic value of life or of some forms of life is threatened by adding new human lives (Attfield 1988)

This argument can only be valid in rather specific contexts. Not only would we have to assume that a limitation of human procreative rights would be the only possible way of preventing a major environmental crisis (a rather eschatological but nonetheless possible history of the world); we would also have to believe that this crisis would be so intense that it would result in the annihilation of all life on earth or in the disappearance of the intrinsically worthy way of life of the Chinese tiger or the Byzantine unicorn.

We can move with more speed now. There is no need to discuss either the validity of the belief in intrinsic value or the validity of the concept itself. The point is that any reference to this notion backfires. If ways of life intrinsically matter, then even a limited restriction of human procreation will result in the annihilation of the human way of life as it used to be. If the number of intrinsically valuable lives mattered, it would be either impossible to justify the sacrifice or prevention of some lives since they are all intrinsically valuable, and if lives are not equally valuable we would still not have solved the rapid reproducers paradox. If life itself mattered intrinsically, it would moreover be impossible to choose between one life and zillions of lives. And, to finish it all off, the notion of intrinsic value cannot help us either to get rid of the objection that any limitation of procreative rights will be biased in favour of the rapid reproducer.

11. *The value of justice is overruled by more serious considerations, i.e. survival and freedom*

We might argue that even if a prohibitive population policy must needs be unjust and unfair and inequitable, there are more important things in life than justice: namely, the preconditions of justice. A just society, it has been argued, would be impossible if there were no freedom, if there were only one possible course of life and no actual opportunity to be unjust; it would be even more impossible if even the resources for that one option had disappeared. Hence, survival is more important than freedom and freedom is more important than justice. (For sources see a.o. Bayles 1976a; Bayles 1976b; Hardin 1976; Lee 1979; Walter 1988.)

Now assume that the obvious rejoinder that survival is impossible without a fair scheme of social cooperation (Hauerwas 1974; Callahan 1976; Hessley 1981; De-Shalit 1996), hence that justice is prior to survival, were empirically true – which seems doubtful. And let us also grant the validity of both arguments. The only result we may expect from a confrontation of the two, apart from a stalemate, is the worst of all possible worlds: if justice is more important than survival, but survival cannot be guaranteed without injustice, the only option that remains is to create a (a bit of) desert and call it justice. We cannot refer now to the necessary injustice of limits to procreative rights, nor hope to refute what basically sounds like a logical argument: no life, no justice. What we can say is that there seems to be no way to morally justify legal limits to procreation, at least not by our moral standard, but that necessity may nevertheless override morality – it may force us to sacrifice a part of our justice to survival.

12. *The importance of procedural justice (here resulting in a liberty to procreate) can be outweighed by that of substantive justice (i.e. the consequences) (Bayles 1976b; Narveson 1973, 1976)*

This final argument might seem to be a less extreme version of the previous one, yet it is not; it is far more radical. Rather than pointing to the necessary requirements for *any* just or unjust, good or evil society, as the previous argument did, this one focuses on, at least, the preconditions of a sustainable society or at worst on the preconditions of a substantively just society. The first thing to note is that there is no such thing as *the* sustainable society (cf. Dobson 1995: 3). Sustainability is a normative, not (just) an empirical concept. Only at the level of mere survival in the most dire circumstances imaginable can sustainability have anything remotely similar to a purely objective, purely empirical meaning – at which point the last argument collapses into the one before. Under all other circumstances, sustainability offers room for choice. Many imaginable worlds are possibly sustainable; the question is which one we find desirable.

To answer this question we would have to do something our stringent interpretation of liberal democracy will not allow us to do: discuss and if necessary dismiss people's ideas of the good life: that is, violate the liberty of life. In particular, it would lead us to question the exchange between procreation and creation: that is, the relative benefits of having children and of preventing a reduction of natural capital, a deterioration of the natural environment or however it be expressed. In other words: we would be forced to say that a plan of life that requires the depletion of our natural resources is not reasonable, hence that individuals' reasons for procreating should be open to public scrutiny and judgement.

I shall return to this issue in the next section, but I can already remark here that the above is a special case of a more general problem: the procedural view of justice inherent in our moral standard is incompatible with the substantive standards that any theorist and proponent of the sustainable or green or Green society would have to set.

6.5 Alternative strategies and parameters

In the previous section, I argued that none of the arguments that are commonly used to defend legal limitations to procreative rights is compatible with a stringent interpretation of liberal democracy. Apart from their individual complications, all face the same problem: that there is no just or equitable way of attributing limits to individual procreative liberties. The balance still tips over towards a liberty for each individual to execute each basic act without limits to time, place, means, ends or intention. The proponent of population policies now faces a choice between three alternatives: to accept less rigid policies, to amend the conception of liberal democracy or to give up hope and opt for alternative solutions to the problem of sustainability. In this section, I shall give a brief survey of the prospects these strategies offer. Up to now, I have assumed that a population policy must address procreation directly by means of legal restrictions and prohibitions. The first amendment that an antinatalist would advise us to make is to adopt an indirect policy instead, i.e. to promote the 'right' rather than prohibit the 'wrong' behaviour. The conventional ways of doing this are either positive or negative: respectively handing out financial and other rewards for small families and families or individuals without children, and denying financial and other advantages to larger families or procreating individuals (cf. Callahan 1976; Walter 1988). Alternatively, we can think of a system of transactable vouchers ('rights to reproduce'). Unfortunately, these solutions meet with the same distributive problem encountered before, albeit in a milder sense: they work out to the advantage of the rapid reproducers and their first children or, in the case of a voucher system, to the advantage of the rich *and* the rapid.

To defend measures of this kind after all, the antinatalist might perhaps find recourse in an argument used by the early Dworkin (Dworkin 1978: 223–65) to defend positive discrimination. The core of that argument is that the advantages and disadvantages which society or the state distributes are meant to guarantee a minimum level of security and welfare, enough to give individuals an equal start in life regardless of their endowments, but not so much as to unfairly (non-neutrally) promote any particular ambitions. On this view, any relative benefits enjoyed by small families and childless individuals would be 'beyond' justice, extras, gifts that governments can hand out at their own discretion. Sadly (for the antinatalist), this argument might perhaps work in the case of positive discrimination (where rejected candidates for a job may have alternatives) but not in the case of a population policy. To name but one serious problem: child-benefit policies of *whatever* kind proposed here are endowment-insensitive and therefore always include effects on children's relative chances in life (*pace* Bayles 1976b).

Note that virtually the same objections apply to the antiprocreationist's next possible suggestion, the use of propaganda against procreation (there is a distinction between propaganda and education – I shall come to that in a moment). Either propaganda does not work, and then it is useless, or it does, in which case it works just like positive and negative incentives with virtually the same effects: effective propaganda is never neutral (*pace* Bayles 1976b). There is a distinction between propaganda and disseminating complete and impartial information, and if the result of public deliberation on this information were a voluntary population decrease, these objections would not apply – provided a government does not itself put population problems on the agenda of public debate. Otherwise, even dispersing objective information can be construed as part of a propaganda policy.

Our moral standard has now brought us back to Garret Hardin's Tragedy of the Commons where, in the eyes of the antinatalist, in themselves perfectly legitimate acts on the part of individuals cumulate in a collective disaster (Bayles 1976b; Hardin 1976), and where, in the eyes of the liberal democrat, there is no morally admissible way of stopping the process. Then if the stringent interpretation of liberal democracy does not enable us to make (some of) these acts illegitimate, the antinatalist's next step can only be that of relaxing the conditions of liberal democracy. Three of these conditions seem to be suited for amendments: the deliberative requirements posed on collective choices, the condition of proportional equality, and the sanctitude of the liberty of life.

In the first place, the antinatalist might argue that any collective choice on population control should take account of the conceivably disastrous effects of un-limited procreation. A full liberty for everyone to procreate would not be prudent, nor responsive to the objective basic needs of individuals, nor good for any common good. Hence, it would be strategically wise to give the conditions posed on collective choice precedence over, rather than let them be subjugated to, the three other criteria of liberal democracy. In that case, we can expect one of two possible results: either a democracy decides, after due consideration, to reinstate the priority of equality and the liberty of life, in which case nothing changes, or it does not, and that would mean throwing away the baby with the bath water, if we may use such an unfortunate expression here. Neither equality nor the liberty of life would be guaranteed any longer, and liberal democracy would be reduced to bare democracy. Even a very limited priority for deliberative democracy opens the door for a dictatorship of the politically correct understanding of the good (sustainable) society, as we saw when discussing the twelfth argument in the previous section. Any further reduction of that same priority would take us back to the eleventh argument again.

A second possible amendment to liberal democracy would adapt the demand for proportional equality in such a way that some differences between children are made morally relevant, relevant enough to outlaw the creation of *j*s but not *i*s. We have, in fact, already considered this possibility to some extent in the previous section. Apparently, there is no logically defensible way to make a significant distinction between the *i*s and *j*s themselves. There may be a relevant difference in the eyes of their parents, though, or at least the antiprocreationist in his innocence might expect that: for parents, every next child may have a smaller marginal utility than the previous one. (Remark that the term utility can be exchanged for other measures of, say, success·in life.) Now assuming this to be true, even for all procreating individuals, and assuming this to be as respectable a view as the Nobel prizes it won, it would still be insufficient to support an argument in favour of legal limitations to procreation. The reasons are simple:

(a) of their own free choice, reasonable individuals for whom the effects of procreation on the environment are relevant will not have *j*s, so that regulations for them are uncalled for (cf. O'Neill 1979)

(b) both they and those reasonable individuals who only care about their own and their immediate offspring's quality of life will appeal to the liberty of life: it is up to them rather than some government to decide when a life may be worth living and when not, or when a society may be called sustainable and when not.

The antinatalist's final option for amending liberal democracy lies in the liberty of life. In view of the previous arguments, he might first suggest that it is not reasonable to (want to) have children, either on balance (weighted against the effects) or in general, so that the desire to have children cannot be part of a reasonable plan of life, so that, in the end, population control and the liberty of life are compatible. Fortunately, we do not need to go into the reasonableness of the reasons people have (for wanting) to procreate.[30] First, if reasonable people procreate, we may – by definition – expect them to have good reasons to do so. Second, if unreasonable people procreate, arguments for limitations to their activities in the field would have to be, at least in part, of a non-environmental nature, i.e. they would have to point to the morality of limiting unreasonable behaviour in general – arguments in which, in turn, the fact that particular environmental circumstances would obtain is no longer a necessary component.

Alternatively, the antinatalist might propose that we replace the idea of a 'reasonable plan of life' by something else – say, welfare, happiness, or 'full plan of life'.[31] But this is no solution either. The whole point of the liberty of life is that it allows individuals to live according to their individual views of the good life, and the precise terms and conceptions with which this concept are expressed should not matter. Any alternative to the reasonable plan of life will either leave it up to the individual to decide what their life should look like, particularly to decide whether children fit in the picture, or it will deform liberal democracy beyond recognition.

In sum, even amended versions of liberal democracy cannot be compatible with legal restrictions to procreative liberties, nor with incentives or propaganda. This leaves the antinatalist with one and only one admissible policy: to appeal, as a citizen, to other citizens' feelings of responsibility. This policy can be supported by one that actually empowers individuals, i.e. that enlarges rather than limits the procreative liberties by making contraceptives and other alternatives to procreation available to those who, as yet, do not actually have the freedom to choose (Walter 1988). Yet it is obvious that no one can guarantee the effectiveness of these policies in ensuring the existence of a sustainable population. Since they are based on voluntary cooperation, volume control is by definition excluded. If we assume individuals to be as rational under these conditions as we assumed they were in the case of compulsory antinatalist policies, we may again expect there to be rapid reproducers who will grab their chances before the limits to population growth have been reached. It seems, then, that the fate of population control is similar to that of other substantive ideals that environmentalists or ecologists may share: the ends-oriented nature of green theories is often incompatible with the process-oriented nature of liberal conceptions of justice.

We return, then, to where we began: to the question with which methods sustainability can be guaranteed. Of the methods mentioned in Sections 6.1 and 6.2, we can discard most so-called demand-side solutions: controlling the length, weight, appetite, diet or taste of individuals, even if practicable, seems even more obviously incompatible with liberal democracy than control of the population size. The latter is acceptable only in the most extreme circumstances, where the survival of humankind is at stake and not the survival of any particular ideal of the sustainable society. Distributive solutions must also be excluded, not because it would be impossible to design and defend a scheme of just distribution and the use of resources within and between generations, but because solutions of this sort (wider distribution of contraceptives, improving the position of women etc.) cannot by themselves *guarantee* that there will never be too many people – or animals, for that matter – to sustain. There may be too many even now to warrant a globally just and sustainable society.

There remain, then, two techniques that may come in handy if voluntary abstinence and distributive solutions turn out to be insufficient: on the demand side, the propagation of austerity, and on the supply side, all technological approaches. Which brings us to the next two chapters.

7

Distributive solutions

7.1 Introduction

In the previous chapter we saw that there is no way in which population policies can guarantee sustainability for liberal democratic societies. The consumer factor cannot be directly controlled. There are too many variables to deal with: immigration and emigration, numbers of deaths and the age of death, people's length and weight, diet and tastes – and the number of births. At the very least the last factor has to be very strictly controlled if we do not want the 'consumer volume' to rise above a sustainable level. However, the only justifiable method of population control seems to be voluntary birth control. From a pessimistic point of view, then, it is impossible to guarantee sustainability. From an optimistic point of view, all we have found is that sustainability through consumer (self-)control is possible – not certain, perhaps not even likely, but still possible.

We may already safely assume that something similar will be true for the supply factor. Since it is assumed that there are limits to growth, we must exclude the possibility of infinite resources. Consequently, there is always a theoretical possibility of scarcity relative to the number of consumers and their needs. The fact that sustainability cannot be 100 per cent guaranteed is, however, no reason to give up all hope. Sustainability is still possible, and if – as assumed – it is also desirable, we are obliged to consider all methods that may bring us closer to sustainability. In this respect, we can expect an independent contribution of distributive justice: that is, independent of the effects of consumer and supply control. A simple illustration: pollution dispersed over a large area will be more easily absorbed by the environment and will do less harm than when it is concentrated in a small area. Distributive solutions can therefore contribute to sustainability, even though they cannot warrant it perfectly either.

In Chapter 5, I sketched a *general* outline of the way in which rights ought to be distributed in liberal democratic societies. I made a distinction between spheres of inalienable rights, unconditional rights and conditional rights. I also developed the restraint principle as an important limit posed on individual ownership rights, as the primary source of limits to environmentally harmful behaviour and as a safeguard of intergenerational justice. Last, I argued for two theses regarding the subjects of rights: for one, that both humans and animals are subjects, albeit the latter in a more restricted sense; for another, that being a subject only means that our interests should be taken into account. It told us little about how subjects should be treated as *objects*. Being recognized as an entity to which rights should be attributed rights does not in any way imply that the entity cannot be used or abused at will for whatever purpose in whatever way.

In this chapter, I take a closer look at the consequences of the restraint principle and our view of rights in general on environmental matters, especially on some of the typically green issues listed in Section 3.5. I do this by increasing the number of variables step by step: first, I consider justice among humans in societies at any given moment of time (Section 7.2), then I add relations with humans outside society ('international' justice, Section 7.3), with other generations (intergenerational justice, Section 7.4) and with non-human nature (interspecies justice, Section 7.5), to finally add a dynamic element: uncertainty about the availability of resources (Section 7.6). Science and technology continually give us three new things: diminishing resources of specific types and new resources of others; new technologies; new information on advantages, risks and availability. In this connection, I want to say a few words about the popular precautionary principle in relation to the restraint principle.

Throughout most of this chapter, I shall make two important assumptions of the 'other things being equal' type. First, I assume there need be no 'objective' discrepancy between demand and supply that would make sustainability physically impossible, i.e. no overpopulation relative to the available resources or no scarcity relative to the existing population. This assumption ensures that we can concentrate on the fundamental issue in distributive justice, 'Who *should* get what?', without being diverted by the empirical question 'Who *can* get what?' The assumption will be dropped, at least in part, in Section 7.6.

Second, I want to improve liberal political philosophy rather than invent the wheel all over again. Thus, I assume that we already *have* a set of liberal principles of distributive justice to work with, principles to which we *add* green elements. Which precise set of principles are presupposed should not need concern us. I want to improve liberalism as such, not any particular liberal theory of justice like Rawls's, Barry's, Ackerman's, Dworkin's or even my own. The improvements I make must therefore serve as amendments to all these theories rather than one specific. The reader who fears this will make the analysis too abstract is free to take, say, Rawls's two principles as a point of reference, but they should bear in mind that I aim at more general conclusions.

7.2 Internal justice

Theories of distributive justice generally set out as theories of social justice, justice in one society, to be broadened in international and intergenerational directions later. I shall deviate from this tradition only in a minor aspect: rather than holding on to the fiction that there are only states and individuals in politics, that states are sovereign and that state and society are congruent, I suggest it is more sensible to discuss distributive justice within the framework of *policy realms*. A policy realm is defined by a political institution with at least one task for which it is exclusively responsible, plus the individuals for whom it fulfils this task (in classical terms: the subjects over whom it has authority). The archetypical state in a Hobbesian world, for example, is responsible for matters of war and peace and answers to no outside authority; as far as defence is concerned, it is a policy realm. On the other hand, local governments are often free in the way they perform certain tasks – they can be allowed, for instance, to construct local roads wherever they like (within their own territory) without being controlled by any outside authority; as far as these roads are concerned, then, the local authority forms a policy realm in its own right.

There are two reasons why I prefer to talk about policy realms. First, it is more realistic. States are no longer sovereign (if ever they were, in historical fact). They are subjected to international treaties on human rights, postal services, measures and weights; to global political institutions that legitimately control what states can do; to regional organizations with far-reaching responsibilities, the most developed of these being (at this moment and to my knowledge) the European Union. In a way, states are often even subjected to lower-level authorities: that is, excluded by constitution or treaty from interfering in the business of policy realms within their own territory – like local governments. More importantly, the primary responsibility for social justice lies with those political institutions whose task it is to distribute benefits and burdens in society – that is, with each of these policy realms rather than the state alone.

I shall assume the physical possibility of sustainability and the existence of an undisclosed set of valid liberal principles for social justice, and discuss only the amendments needed in order to further sustainability. What follows logically from a commitment to liberalism and the liberty of life, is that I understand sustainability as social sustainability: life must not only be protected against extinction and scarcity, it must also be endurable for those creatures that can experience it. Hence, I shall deal with policy realms where natural resources are scarce only in a minimal economic sense, i.e. where there is a constant flow or stock of resources and more than one possible owner (competitor) for each unit of resources. My question now is: what happens when we introduce the restraint principle and the distinction between conditional and unconditional rights in such a policy realm? In this section, I follow the 'natural' development of goods from nature to resource, product, trade, consumer good, waste product and finally back to nature step by step; in the next sections, I can afford to be less detailed – in this one, I hope to be *very* thorough indeed.

Nature must be transformed into natural resources before anything else can follow. This involves basic acts, not of a physical but of a psychological nature: we have to look at a cow not as part of the animal republic but as an object that can be transformed into milk, butter, cheese, fertiliser, meat, glue, dog food, coats, pants, purses, ties and EU subsidies. There is no difference between humans and animals here. Basically, this is the way the lion looks at the antelope around lunch time. More importantly, in the cause of social sustainability, these basic acts cannot be wrong in themselves. For life to be worth living, in liberal terms, for the individuals' liberty of life to be promoted, we need resources. Although not every separate unit of resources has to be a necessary condition for survival and social sustainability, it *can* be one and the category resources as such certainly is. It follows – and here comes an abstract formula – that any creature to which rights can be attributed because the liberty of life, or consciousness or agency, means something to it, i.e. any 'subject', deserves the unconditional right (freedom) to consider the objects around it as means to satisfy its interests in consciousness, agency or the liberty of life. A formula like this is perhaps too labyrinthine to be grasped at once and certainly too complicated to be repeated. As of now, I shall economize on language and concepts by identifying 'subjects' with humans only. It follows then, in short, that humans have an unconditional right to consider nature as a possible resource. No restraint is required.

Next, the things we can think of as resources must be physically transformed into resources. We have to distinguish three cases: some specific units of resources can be vital to the liberty of life of an individual, some can be vital for social sustainability, and some can be neither. If my life means nothing to me without ownership of the *Mona Lisa*, the first case obtains. If without me, humankind cannot survive, the second case obtains (and if I need the *Mona Lisa* like humankind needs me, both obtain). If there is no one who cannot live without the *Mona Lisa* and if there is no other reason why humankind cannot survive without it,[32] the third case obtains. Our first case is the basis for a classic dilemma in justice theory: suppose that there are two people lost in the desert, and they only have enough water left to save the life of one of them – who shall it be? Fortunately, it is possible to circumvent most questions of this type. I have already assumed the existence of principles of justice which will undoubtedly give some kind of guidance in choices of life and death. Green considerations cannot add anything in a case where more than one claimant qualifies for an unconditional right to a resource, unless it is the fact that they may have given an independent reason to distinguish between conditional and unconditional rights. Nor can they add much to a decision when a choice has to be made between a person for whom *x* is a basic need and another from whom *x* is merely desirable; existing liberal theories of justice already make this distinction.

Since I am talking about humans exclusively in this section, there is only one situation in which green ideas *can* be of assistance: in the most difficult problem of all, when the resource in question is itself a subject – say, when a plane crashes and the only way to survive is if some of the survivors eat some other survivors.

In all other cases where legitimate claims to unconditional rights 'outnumber' the available resources, typically green ideas that might be helpful will be inadmissible. They would require that we compare plans of life by their 'environmental friendliness', which cannot be done without making unconditional rights conditional, both in a particular case and hence in principle, thereby denying the liberty of life. Yet where two or more subjects are themselves resources, we can consider the relative merits of the resources as resources and so add information that may, repeat, may, end the deadlock. It could help to know – to stick to environmental arguments – if Peter's survival is a necessary condition for the liberty of life of a third party, or more important than Paul's survival, or that either one's survival is (more) vital to social sustainability. However, it can also be argued that once unconditional rights have been established (not to mention Peter and Paul's inalienable rights to certain bodily parts), there is nothing more to be said from a (social) justice perspective. The acceptance of further decision criteria would be a matter of generosity, benevolence, love, and at any rate an ethical choice to be made by the individuals among themselves, since liberal political institutions are not supposed to judge individual conceptions of the good.

Our second case involved the transformation of nature into resources that are vital to social sustainability. This is an easy case. No one who values sustainability can reasonably object to actions necessary for sustainability. Yet even here, I can create a few problems. First, what may be vital for social sustainability need not be vital for economic or ecological sustainability, let alone for more deeply green views on the relation between humans and nature. Since I am considering things from a liberal point of view, this need not concern us too much: it is social sustainability I care about most, and social sustainability should not be sacrificed to any lesser ideal. Still, it is proper that we remind ourselves of the limitations that a commitment to social sustainability puts on our love as much as our dislike of nature.

The second problem is this: what if I have an unconditional right to x but want to use it in a way that is not compatible with the vital contribution x could make to social sustainability? In this situation, either claims of other members of present generations are at stake, claims that will indubitably qualify them for unconditional rights that are at odds with mine, or there is a contradiction between my unconditional right to x and the unconditional rights of non-existent future generations or other subjects of rights. The latter is a matter for Sections 7.4 and 7.5, the former typically an instance of our first case, where the unconditional rights outnumber the available resources. Last, let us consider resources that are neither vital to individuals nor to social sustainability. To picture this case, imagine a country with two sweet water lakes, each of them containing enough water to ensure the survival of the entire population until the next rainy season. The population then has an unconditional right to the amount of water contained in one lake, but neither one of these lakes is *specifically* necessary for the survival of our population. It is here and only here that the restraint principle comes into force – full force. What the restraint principle demands is that each single transformation of nature into

163

resources ought to be justified rather than be assumed to be legitimate unless proof to the contrary is offered. In our example, we would need reasons, selection criteria, to justify why so-and-so much water should be taken from one lake and not from the other. What the restraint principle also demands is that anything that is taken out of nature should somehow be replaced. Both demands will turn out to be relevant in this example.

Since there is no way of distinguishing between 'water mixes', between taking 100 per cent or less from one lake and 100 per cent or less from another as long as the total covers the full needs of the people, we can only look at the effects of each possible act on the lakes' environment. In doing this, we have three criteria. First, it is possible that totally depleting one of the lakes or depleting it to a certain degree (call this: Situation 1) would indirectly harm, say, Evelyn, in her *unconditional rights* to things other than sweet water – for instance, and to oversimplify the case, because the lake's banks would dry out, the plants there would die, so that her farm animals and ultimately Evelyn herself would starve. A conflict between unconditional rights gives us a prima facie argument against creating Situation 1.

In Situation 2 it is possible that there are no unconditional rights violated by any water mix – not by taking 100 per cent or less from one lake and not by taking 100 per cent from the other. Here, people are not prevented from getting their basic needs but they may be hindered in the (further) pursuit of their plan of life. The liberal principles of justice already instated will, we may assume, ensure that the best (most impartial) water mix is chosen, so that as few persons as possible lose as few conditional rights as possible. The green amendment to these principles would want us to take environmental effects into consideration, i.e. the numbers of basic acts (conditional rights) that are excluded by each water mix directly and indirectly due to environmental degradation *and*, since the restraint principle demands that we replace anything we take out of nature, the costs (or even impossibility) of restoring the damage. In other words, our amendment would ask for a kind of cost-benefit analysis measured in freedom and basic acts rather than money.

In Situation 3, every possible water mix violates someone's unconditional rights in one way or another. There are thus prima facie reasons against both creating and not creating Situation 3 – a deadlock. Here, we have a mix of two problems. On the one hand, it is an instance of the classic dilemma of legitimate claims to unconditional rights outnumbering the available resources described (and solved, in so far as possible) above. On the other hand, the problem is similar to that described in the previous paragraph. It is impossible to leave rights to nature, in this case unconditional rights to nature, untouched: we will again have to look for replacements of the lake, for close enough alternatives or ultimately for ways of compensating those deprived of their rights. Assuming that we can actually choose who gets the indispensable resources and who will not, the solution is the same: a cost-benefit analysis in terms of freedom.

Resources, once created, will be further transformed into products, be traded and become consumer goods. There is no need to be concerned with the effects of the restraint principle in these phases because there are none, at least no new ones. Transforming a resource into a product and using it as a consumer good is, for all theoretical purposes, virtually the same as transforming nature into a consumer good. Relative to others who do not own or use a specific product or consumer good x, x is comparable to nature: for them, it is a good that can (physically) be taken away, transformed into something else and used – the sole but inconsequential difference being that x is artificial in nature. Trade, moreover, is comparable to the very first phase I described: that of psychologically transforming nature into resources. In none of these cases would it serve any use to repeat the analysis.

Some readers may find it objectionable that I seemingly allow any good to be traded, but I suspect there are no grounds for objections of that sort. The comparison between trade and 'thinking resources' merely establishes that there is a right to 'think trade' under the same conditions: that is, in the pursuit of sustainability and the liberty of life. This does not imply that any good can be traded any time for any reason or no apparent reason. It specifically does not imply that if our calling in life is of a commercial nature, we have the right to trade any specific good or type of good. To trade as a *physical* activity, the same restrictions apply as to physically transforming nature into resources – *mutatis mutandis*.

Moreover, critics of a more liberal inclination might object precisely to these restrictions as unnecessarily or unjustifiably limiting the free market, whereas more green critics might object that it leaves too much room for industrialism and all the environmental evils it has caused in its incarnations as communism, capitalism and imperialism. The first of these two objections seems to assume that there are no moral limits to free trade, that if there are any limits, these cannot be environmental in nature, or if they are of an environmental nature, they cannot be implied by the restraint principle. The second partly confuses the free market with latter-day capitalism and partly confuses classic economic scarcity with morally relevant scarcity (i.e. 'demand' with 'needs and wants'). Apart from being mislead or misleading, these objections are also misplaced: they relate to the best *way* of allocating goods and resources rather than the primary observation that goods can be allocated. Whether trade and exchange in some form or other are morally justifiable is a totally different question from the one about the proper way of promoting fair trade in the real world. I shall therefore postpone a discussion of these objections until Chapter 9.

There is one last aspect of the consumption cycle that needs attention here: waste. Both in the development of goods from nature to resources, products and consumer goods and at the end of the line, after consumption has taken place, we produce waste and return it to nature – sometimes with detrimental effects (a.k.a. pollution) on nature and sustainability. Precisely because it has environmental effects, waste is neither an environmentally nor a morally neutral category. Its types and amounts may limit the individual's formal rights (in everyday language and libertarian terms (cf. Nozick 1974: 78 ff.): *violate* rights) and influence their life prospects.

On this perspective, waste is the mirror image of resources: whereas resources further the individual liberty of life and secure overall social sustainability, waste diminishes and jeopardizes them. In so far as resources are goods, waste is an 'ungood'.

The ideas that gave birth to the restraint principle can, for the very same reasons, give birth to an inverse restraint principle for waste. I introduced the restraint principle because rights to perform basic acts must be justified separately and because there is no argument other than need that can unreservedly justify the destruction of a good. If no good may be destroyed (unless necessary) then no good may be wasted (unless necessary): in other words, no ungood may be produced (unless necessary). Substitution of the terms relating to goods in the restraint principle by more appropriate terms gives the following *inverse restraint principle*:

> no waste shall be produced unless unavoidable and unless whatever is wasted is returned to nature without degrading it in any way; if that is physically impossible, nature should be restored to resemble its original state as closely as possible; and if that is also impossible, a proper compensation should be provided.

Thus, there is no moral objection, for example, to burning coal as long as (a) burning coal is necessary in terms of the liberty of life or social sustainability; as long as (b) the necessary mining activities and the resulting pollution are spread over an area large enough to allow nature to absorb it, or in so far as this is impossible; as long as (c) both the land (surface and anything beneath it) are restored afterwards, or in so far as this is also impossible; as long as (d) alternative measures are taken to maintain social sustainability and compensate those whose liberty of life is reduced by the effects of mining and pollution.

As the example may show, this is a principle with on the one hand apparently rather rigid and exacting, almost Utopian implications. In combination with the original restraint principle, it seems to demand that we look on the relation between humankind and nature not as a relation of exploitation limited only by the sky, nor even as one of producing, consuming and recycling *goods*, but as one of recycling *nature* and subjecting human desires to that end. On the other hand, is seems to be a principle that offers more than enough room to exploit nature to the fullest extent possible, provided the right reasons are found.

To start with the latter point, the inverse restraint principle adds nothing essential to the effects of the restraint principle. Exploitation of nature still has to be justified in terms of basic needs and further wants and the satisfaction of the latter is still limited by the protection of the former. Whether these are purely formal restrictions on exploitation with little or no effect in real life must remain open to debate until the international, intergenerational and interspecies dimensions have been added to the picture – that is, until Section 8.1.

Yet it is true that waste and pollution are here to stay as long as we value the conservation of nature in the state that it is in at any given moment in time. (The presence of humans is not required: it suffices that something occurs that irreversibly changes an ecosystem.)[33] It is also true that nature often cannot be returned to its previous state or to one closely resembling it; compensation and not restoration may sometimes be the only possible answer to exploitation.

To make matters worse, the demand for compensation may not be as much of a deterrent as it is sometimes hoped. It could be if the costs of compensation were distributed in accordance with the famous 'the polluter pays' principle, and if the polluter were unambiguously identifiable, but there is virtually no way in which this principle can be added as an amendment to the liberal principles of justice I postulated. It is unfair to punish someone who, in order to survive, has to cook a meal, burn wood and leave waste behind; in general, and I admit there may be exceptions, it is unfair to punish an individual polluter who has no choice but to pollute – just as it is unfair to punish anyone who kills in pure self-defence. It is also unfair to punish the one producer of waste among many people who could have been in their place if what they produce is vital to social sustainability or individuals' liberty of life. Finally, punishing the (immediate) polluter can be unfair even if the polluter's activities are unessential to either of these goals: we do not punish the murder weapon but the murderer. 'The polluter pays' seems justified only when there is a real choice between polluting and not polluting, i.e. a choice other than that between sustainability or liberty of life on the one hand and perishing on the other. Where there is no such choice, the 'polluter' cannot be said to benefit. It is more just, then, to demand that the beneficiary pays.[34] The problem with this demand is that usually many people if not a whole society profit from waste-producing activities. The more beneficiaries there are, the less their individual share in the total costs of compensation will be – hence, the less deterring a demand for compensation will be.

As to the reverse objection that the two restraint principles may be too demanding, there are two possible answers. The first is a traditional green one: there is no such thing as a free lunch. With or without restraint principles, there are costs attached to waste and pollution, and these costs are bound to pop up sometime somewhere. Hence, it is better to be aware of them and account for them in the (abstract) form of principles like these and in the (concrete) shape of legal and financial duties. Up to the conclusion, this is a quite reasonable argument – only the conclusion does not follow. The free lunch argument merely points out that someone sometime somewhere will bear the consequences of waste-producing activities; it gives no reason why the costs should be paid here and now. We can patch up the free lunch argument by adding the arguments given in favour of both principles and thus save the conclusion, but that will only serve to show that it is just to pay the costs of pollution immediately. It still will not answer the objection that the restraint principles may be too just, too demanding: *fiat justitia pereat mundus*.

The second answer is essentially evasive. It would admit that the restraint principles limit the freedom of action and lifestyle of some citizens (the 'wasters') more than that of others (call them the eco-vego-burghers) and thus create inequality; it would argue, in a 'free lunch' style, that these unfreedoms and inequalities are justified. But it would also point to the fact that the restraint principles are in fact liberal principles, designed to ensure that individuals get the fairest possible chance to accomplish their plans of life, whatever their plans may be. The restraint principles are there to promote the satisfaction of basic needs and the protection of further wants. It limits no desires except the desire to destroy, and even in that respect it is moderate enough to allow an occasional senseless, Dionysian spree of vandalism and demolition (an admittedly *very* sober example: the traditional Russian vodka toast) since even a fair bit of ridiculous and 'irrational' behaviour may be a basic need, a component of the full life. Rejecting the restraint principles as too demanding implies that liberalism itself would be too demanding. Whether it is – and this is where I feel a need for evasive action – remains to be seen once our picture of green and liberal distributive justice is complete.

One of the aims of this section was to show that environmental costs and benefits can and should be treated like all other costs and benefits. I want to end with an illustration of how the restraint principles might actually influence the distribution of burdens and benefits in society, merely to show that these principles *can* have some effect without becoming too demanding. I assume once more that pollution and compensation are here to stay and that 'the beneficiary pays' is a just principle for the attribution of the social costs of pollution. Under these conditions, it seems fair to demand that the costs of pollution 'travel along' with the benefits, i.e. from resources to consumer good, until they reach the ultimate beneficiary. One way of doing this would be to introduce a tax that works in much the same way as VAT does in some countries: a Pollution Added Tax, PAT. It would oblige the first owner of a resource to charge their clients a percentage of the price for pollution costs, which would then be transferred to a collective fund (Treasury, Environmental Agency, whatever). Assuming that they alter the good (and in doing so produce more waste) and/or that selling goods may itself involve pollution, the second owner adds the same percentage, passes the (cumulative) charges on to the second buyer and pays the difference between the first and second sale as tax, and so on to the last consumer. The catch is that there must be a way in which PAT money can be used either to clean the environment or to compensate individuals for the degradation of their bit of the environment.

7.3 'International' justice

In the previous section I studied the implications of the restraint principle(s) on the distribution of benefits and burdens within policy realms; in this section I cross the borders between policy realms. So far, the impact of introducing policy realms in place of sovereign states was hardly noticeable. Things change when we consider relations between 'states' and 'realms'. Policy realms can be cities as well as federations of states and anything in between. To anyone who is accustomed to the idea that justice between two neighbouring local communities is of a totally different order than genuine international justice, the introduction of policy realms must sound like a prelude to confusion. After all, do not neighbours have obligations to one another that strangers meeting in the desert do not?

I have no intention of contradicting this common sense objection, but I would like to add a few nuances. Rather than making a sharp distinction between national and international justice or totally denying any moral difference between territories of justice, I suggest that we ask why we, or some of us at least, would want to make a distinction between our obligations to our fellow citizens and strangers in faraway places. The answer I formulate leads to a view on 'international' justice that allows for more subtle degrees of concern.

The following is one possible description among many of the relations between individual humans and/or policy realms. I do not intend to defend it, because any other description would, or should, serve equally well to bring out both the morally relevant difference and the similarity between 'embedded' and 'unencumbered' relations. When two people meet as strangers in the desert and when at least one of them could benefit *from* or be harmed *by* the other to the benefit or disadvantage *of* the other, they become mutually responsible for one another. The more possible relations they have, the more responsibility they carry; the more these possible relations are translated into interactions, the more people become involved in one another's lives; and the more people get entangled in such relations, the more complex their network becomes. Some of these relations can involve 'iterated' interactions between two or more people interacting with, in principle, the same frequency (e.g. love) or between one constant factor and incidental others (e.g. trade). For these relations and for the goods exchanged therein, customs and rules develop: individuals thus become embedded in social spheres. At the same time, rules and customs create expectations about the way in which future interactions will take place. Individuals partaking in social spheres become morally responsible for breaking or upholding the rules and for the effects their actions have on others. The social sphere itself, then, becomes a kind of common venture. Where social spheres (the networks, the individuals *and* the rules) become congruent, a (civil) society exists; where they are covered by political institutions, they become policy realms; where and in so far as societies are covered by congruent political institutions, they become states. On the other hand, the less people are embedded in social spheres or embedded in less social spheres, the more they are strangers to one another.

Imagine two thirsty couples, one member of each couple having enough water for both. Neither couple, I assume, has ever before been in comparable circumstances. The first couple, two strangers meeting in the desert, is unencumbered: they have no prior history of interactions, they do not participate in the same social spheres or policy realms. (Concluding otherwise would be committing the naturalistic fallacy.) The second couple, two neighbours meeting in the street, are on the other hand usually embedded in a society: up to a point, they share the same history, language, culture, tastes, rules, interests. The two strangers are empirically responsible for one another's fate but need not be morally or rationally obligated to accept their responsibility. The two neighbours, however, are both empirically and normatively responsible. The neighbours are involved in a common venture: for them, there is more at stake than an incidental sip of water. Other things being equal, i.e. the two not being personal enemies or anything of that sort, a decision not to share the water will damage their existing relationship and at the very least hinder future interactions. It is easy to extend the case of the two neighbours to larger sets of individuals and other (types of) interaction(s), the result being that the stakes grow ever higher. In the end, too frequent or too obvious or too fundamental instances of not accepting one's moral responsibility will damage the whole fabric of society.

The difference between our neighbours and our strangers, then, lies in the existence of established responsibility for one another's fate: in the case of the strangers, the existence of moral responsibility still has to be proven. Yet there is also one important similarity: in both cases, we are talking about a specific, new type of unencumbered interaction, i.e. one for which rules do not (yet) exist. The fact that our neighbours *are* and that our strangers *can be* morally responsible for one another does not give us any clue as to *how* their (hypothetical) responsibility should be given shape. In this particular sense, even neighbours are strangers.

Let us translate these conclusions to the level of responsibilities between policy realms and their mutual environmental fate. The mere possibility of interactions that influence the environment of other communities creates empirical responsibility. Apart from the possibility that prior relations exist, there is no reason to believe, nor to disbelieve, that this empirical responsibility implies moral or rational responsibility. And all in all, there is no reason to believe that a possible normative responsibility should be given any particular shape. The question then is, roughly: how should we care about some obscure tribe in some dark corner of the world that sees other obscure individuals enter their territory and cultivate it in ways incompatible with their way of life, perhaps even threatening their conditions for survival – if we should care at all? We do not have to take a roundabout way past more fundamental issues to answer the latter question. I assume once more that standard liberal principles of justice already offer satisfactory solutions, whether affirmative or negative, to demands like those for the rectification of past injustices (colonialism, slavery, war) and for the rectification of the current undeserved inequality between North and South. What we need to find out is how environmental issues would relate to these principles.

There are two cases to be distinguished here. In the first, environmental conditions threaten or further the interests, needs and wants *only* of people in a policy realm distinct from ours. Let us assume that it is a country and call it Arcadia. Here, outside interference of whatever form – aid, trade or participation in gains – is purely voluntary. Whether it would be legitimate to interfere seems to depend on whether the Arcadians invited us to or rather, since there is an analytical distinction between expressed and true preferences, on whether they should in all reason have invited us. If they do not or should not, then we have no business there; if they do or should, then they have created an (empirical) responsibility for their and our well-being and our first case collapses into the second. In the second case, both our and their (environmental) interests are involved. This can mean anything, from colonizing Arcadia and solving our own environmental problems by exploiting their natural resources, to being invited to visit Arcadia's health spas, to being endangered by Arcadia's dumping nuclear waste in the river from which we drink, and from singlehandedly solving Arcadia's environmental problems, to committing all sorts of sins of omission.

A case can be made for our getting involved in Arcadia's affairs as if they were our own, on the basis of the moral relevancy of the Arcadians. For this, it is irrelevant whether we like the Arcadians and their way of life or not. Both the age-old romantic love for the noble savage (to which most recently several tribes of Amazon Indians have fallen prey) and the probably equally old aversion to the short, nasty and brutish life of barbaric primitives are a matter of purely personal taste. From the point of view of a liberal democracy, the design and development of people's plans of life and theories of the good life are certainly not any government's business. Nor should it be relevant that the Arcadians have a different nationality. Nationality is as deserved or undeserved by an individual as any other artificial endowment, including the (dis)advantages attached to innate endowments or impairments (cf. Barry & Goodin 1992; Carens 1992; Steiner 1992). Consequently, the differences that come with nationality, the economic and social advantages one individual gets through birth and a second, being 'citizenship-exploited' by their birth in a less prosperous society, are equally arbitrary (cf. Van Parijs 1992). What is relevant – what makes the Arcadians morally relevant – is that we are in a position in which our political institutions by attributing rights can further or impede, even destroy, the way of life and the liberty of life of individual human beings as *persons*. Persons are morally relevant by definition and not because of any accidental quality like race, creed, condition, sex or nationality. Being in a position where rights are *attributed* to persons, a political institution is as morally responsible for the stranger as it is for the neighbour.

Liberal principles of justice may demand that more is made of this responsibility with regard to 'embedded' neighbours than to 'unencumbered' strangers because the institutions of a policy realm have more and deeper obligations to the former. Hence, being a stranger may (but need not) be a decisive factor in the kind of life-and-death choices described above in Section 7.2. Aside from this, strangers and neighbours deserve the same consideration. There being no further relevant difference between them, we can apply the rest of the analysis in Section 7.2 without any difficulty to the case of interrealm, 'international', justice – that is to say, we may conclude that green considerations add nothing except for the restraint principle to existing liberal principles of justice.

However, the fact that environmental ideas add so little to existing liberal theories of justice should not be mistaken to mean that they also add little in practice. Quite the contrary: in so far as a moral principle can influence everyday life, the restraint principle has very profound effects. It supports the green conviction that it *is* fundamentally unjust to, say, exploit tropical forests by simply cutting them down: it involves destroying the environment of 'primitive' peoples, destroying their way of life and critically diminishing their liberty of life. It also supports the conviction that 'foreign' political institutions can have a partial moral responsibility for the fate of tropical forests and faraway people. The restraint principle would demand, first, that forests are not cut down permanently (e.g. for agricultural purposes) unless there are no less harmful alternatives; that if they must be cut down for wood, it should be done in a way that least burdens the existing local ecosystem's capacity for recovery; that if an entire area must nevertheless be cut down, the forest be replanted; and that anyone who must involuntarily suffer negative consequences should be compensated in another way than by leaving them an uprooted society, barren land and an empty stomach. The restraint principle would demand, second, that no tree be felled to satisfy foreign desires for tropical hardwood if there are alternatives to this type of wood; that in principle (i.e. apart from life-and-death choices) no forest be cut down for this purpose if it involves violating the unconditional rights of local people; that the buyers and users of tropical wood pay for the waste they cause or allow to be caused and compensate those whom they harm.

Then again, it should be noted that the restraint principle can be a two-edged sword. As discussed in Section 7.2, it is consistent with the psychological transformation of nature into resources: there is no prima facie reason why any bit of pristine nature should not be considered as a possible consumer good. It is, furthermore, as consistent with the restraint principle to allow foreign consumers to (at the very least) criticize the environmentally hazardous activities of producer countries as it is to criticize them for protecting nature 'too much'. Both involve the denial of exclusive territorial rights to the unconditional ownership and use of nature. Also, the restraint principle supports a more equal access to natural resources, clean technology and so forth as much as it can support a global redistribution of waste and pollution. It is, after all, a distributive principle and not one describing a most desirable end-state.

Finally, one effect of amending liberal principles of justice with green demands is especially worth mentioning: its influence on sustainability under imperfect political conditions. Political institutions do not always operate as, or within, a well-ordered structure but instead often in a hostile environment, where hostility can take any form from social and political division to outright civil war. A modern and modest example is Russia, with its many competing official agencies and authorities as well as semi-legal and illegal intruders on the market for political 'services' (cf. Weenink 1997). More extreme examples are countries like Ruanda, Burundi, Ethiopia, Somalia, Afghanistan, Cambodia and so forth, where at one time or other no government worth mentioning was left. It has been noted (cf. Neefjes 1996) that many of the conflicts that result in the disintegration of societies and political structures originate, at least in part, in environmental problems: disputes over scarce *natural* resources. Let us assume that these are instances where the parties involved should reasonably be expected to ask for foreign help or where foreign parties already have ties and responsibilities. Let us furthermore assume that it is actually possible to (re)create the conditions under which a society can be revitalized and where the (environmental and other) requirements for social justice can be met.

In these initially hostile circumstances, social sustainability (that is, the survival of numerous lives worth living) may require a lower degree of satisfaction of the restraint principle than we would expect in 'ordinary' societies. The *technical* solution to the environmental problems at the background of such deep social conflicts could lie in a more efficient use of natural resources. This demands coexistence and cooperation: Hutus and Tutsis living even closer together, reallocating land to farmers and rangers, using the land more efficiently, or Russian industries and authorities promoting energy-saving technology. But it is probably too much to ask that two peoples who only a while ago were involved in mutual genocide should live together and cooperate, or that organizations and companies that could be better off in the future *if* they survive innovation should immediately jump into the crevice. The *feasible* solution can be far less environmentally benign: it may demand an increase, even a huge one, in the extent to which nature is exploited, the land cultivated and resources spent. In terms of social sustainability, this is an investment in the future. It is the kind of sacrifice of nature that must be accepted to facilitate long-term survival. In terms of environmental sustainability, it may well be a disaster for non-human nature. There are limits, then, to how green green liberalism can be: it will not sacrifice human lives to any calf of any colour including gold.

7.4 Intergenerational justice

I began my analysis of the possible contribution of distributive justice to social sustainability with an (almost) ordinary well-ordered society and then added 'international' justice. To complicate matters further, I now consider the claims and interests of other generations, past and future.

In line with what has been said in Chapter 5, I must make the realistic assumption that 'generations' do not really exist, and certainly not as sets of individuals who mysteriously appear and take the place of a previous set which at that same moment mysteriously disappears. People, individuals, of different ages coexist, some are born, some die, but no one takes anyone else's exact 'place'. It is possible to refer to individuals of roughly the same age as a generation, but only if we keep in mind that the term is just shorthand and a metaphor. It can also be used by an individual to distinguish the children and grandchildren (etc.) of her parents and grandparents (etc.), but it is even less helpful to us in this sense.

Justice between existing 'generations' is a necessary condition for the stability of a just liberal democratic society. It also turned out to be possible to defend the restraint principle both as a valid interpretation of John Rawls's savings principle for justice between generations and as a ecumenically valid criterion for liberal justice between generations. Both in Chapter 5 and in the preceding sections, I considered the effects of the restraint principle on existing persons. In Chapter 5, I inferred that the restraint principle also implicitly protects the interests of non-existing generations in that it will affect them in the same way as it affects the younger 'generations' of existing people. Here, I want to ask if acting in accordance with the restraint principle is *all* we owe to non-existing individuals, or if there is perhaps more that should be done.

We should first distinguish between two types of non-existing individuals: those who preceded us and those who may come after us. Past and now extinct generations are no longer of this earth but off it; they have no interests any more, no plans of life, they cannot be harmed nor benefited. When meting out distributive justice, we cannot give them anything nor do we owe them anything. Still, it could be argued that individuals may be obliged to keep whatever voluntary promises they made to the dead while they were alive as much as they are obliged to keep promises made to people who are presently absent for other reasons, since a promise is an obligation to ourselves as well as to another. By the same token, a society of existing individuals may, as a whole, be obliged to keep a promise made to or fulfil a task imposed by its dead predecessors. However, this makes whatever is 'owed' to previous generations part of the plan of life of existing generations: whatever we owe them, we owe ourselves. In matters of distributive justice, then, the notion of (obligations to) previous generations is redundant.

Non-existing future generations are a totally different matter. Provided they come into existence, we can and do harm or benefit them, and depending among other things on how we estimate that we influence their lives, we can decide to bring them into existence. Unlike the dead, we are responsible for their fate. Now there is no legitimate way in which a liberal democracy can regulate the number of children present persons conceive and give birth to, nor, by implication, the distribution of births (cf. Chapter 6). Hence, even if we were able to satisfy the restraint principle as strictly as possible and even if we knew what their preferences would be, there is no way in which a liberal democracy can *assure* that future generations will not be worse off relative to present generations. Regardless of whether we care for future generations at a 'discount rate' (i.e. more for the next generation than for subsequent ones) or whether we are temporally impartial (i.e. care equally about all future generations), we should care *at least* about the child we personally create and *at least* in a negative sense: we should not make its life unbearable (cf. Narveson 1973). After all, by creating it, we create the circumstances in which it will have to live. Being empirically responsible for its existence, we become morally responsible for the options it has to live a tolerable life. It may be possible to argue that we owe more: more than just minimum conditions for survival at an endurable level of well-being, more than just minimum conditions for the children we personally create, more than just this for just the next generation. However, even this minimal argument suffices to establish that procreating individuals *can* owe more to their children than the restraint principle demands: the restraint principle is an open-ended principle and as such allows that there may just be too few resources left to ensure that a hypothetical next ('post-last') generation can lead a tolerable life.

If individuals can owe more to their children than the restraint principle demands, will not their neighbours and their society as a whole then also owe more to the next generation? The answer, I shall argue, is that we do, sometimes. We do not necessarily owe them more resources than the restraint principle demands, nor are necessarily obliged to make our lives (even) less comfortable than they are to benefit them, but we do owe more than mere restraint.

There are many significant differences between actual and possible persons that make it difficult to see the latter as somehow similar to the former in a morally relevant way. We do not know which possible persons will be born to whom, we do not know what they will look like, what their abilities and disabilities will be, what preferences, plans of life and theories of the good they will have. Most importantly and preceding all these uncertainties is the most fundamental uncertainty of all: unlike us who already exist, their existence is contingent. And this makes a tremendous difference. The creation of a specific child with specific abilities and a specific plan of life is, notwithstanding significant advances in genetic engineering and social conditioning, still for the greater part a matter of fortune, an involuntary business. The creation of a child as such, on the other hand, is never a necessity (an implication of a law of nature) and it can be a voluntary act. Let us assume first that it is. And let us assume – since I already assumed a society where goods are only moderately scarce – that adding a child to an existing population will not push anyone below the level of a life worth living but that its existence is a net loss for others: there will be one more person to share in the available, constant, amount of resources. Obviously, if a new child is *not* a net cost for others, that would be because it will contribute at least as much in resources etc. as it will use – so there would be no need to save more than the restraint principle demands, nor to discuss the theme.

And now, a child is born. At the level of individuals,[35] I have reason to complain about your causing this child to exist: by doing so, you deprive me of a part of my share in the world and with that of part of my freedom, part of my options, part of my dreams. You, on the other hand, could give at least two answers. First, you could say that your child has as much right to its share in the available resources as you or I have – does it not bleed? and all that. This would typically be the wrong answer: it does not justify your voluntary choice to (create this child and) deprive me of part of what I could reasonably presume to be my rights. Next, you could reply that the only way in which you could not 'harm' me would have been to not have this child. Yet that would, you argue, mean that you would have had to harm yourself more than you now harm me: having a child means more to you than this miserable speck of resources you deprive me of can ever mean to me. We need not be libertarians to admit that there is something basically wrong with harming others for whatever personal gain, unless it is a matter of life and death – or in the context of this book, of a tolerable versus an insufferable life. You will thus have to modify your answer and claim that having a child is a vital interest for you, a basic need. In that case, and only in that case, can it be argued that I and with me the rest of society owes more to possible future persons than restraint alone: a redistribution of our rights to reserve some extra goods for future generations.

Suppose next that a child is born and that its creation was *in*voluntary. (To avoid a further difficult issue, that of abortion, I assume that its birth is by implication also involuntary.) There are dozens of reasons why this could happen and with that many degrees of involuntariness: from rape to the non-availability of contraceptives, from ignorance and accidents in bed to lack of backbone, from cultural pressure to conform to the 'rules' of our gender and peer pressure to insanity and lack of control over our lust. The point is that, in so far as this child was created involuntarily, its parents cannot have been obliged in advance to save on its behalf – no more than anyone is obliged to save money for a robber or a burglar who may one day visit us. That individuals *qua* individuals are not obliged to prepare for one particular accident does not, however, imply that society cannot be morally obliged to save resources for this or any accident. Accidents happen to all of us at one time or another, and if they do not, they could have happened to us – that is the definition of an accident. The welfare state, one of the guises in which really existing liberal democratic societies present themselves, is built partly on the idea that it is fair and just to pool the risks and share the costs of accidents, regardless of whether these risks are natural like earthquakes and floods or artificial like traffic and work (Ewald 1986). One argument in support of collective insurances is that since an accident could happen to anyone it is unfair that it hits one specific person and no others; it would be more fair if it hit all of us a bit. Alternatively, collective insurance can be seen as a good thing because it prevents the victims of accidents from becoming a burden on individuals and charity.

We do not have to be libertarians to object to such schemes. First, both arguments are inconclusive. The premise that no one deserves to be hit by an accident is incompatible with any argument that results in the conclusion that we all deserve to be hit by one, just a bit. The idea that victims of accidents would be burdens on others confuses justice and charity. If it is a just act to compensate accident victims by means of an insurance scheme, it is also a just act to compensate them in other ways; the objection to other schemes must lie in the injustice of the *method* of compensation. But the alternative method is charity, and charity, even if it is a moral obligation rather than a freedom, is not analysable in the burdens-and-benefits vocabulary of justice. Besides, as a libertarian would argue, how I direct my life is my affair as long as I do not violate the rights of others. If I decide to run risks, then the costs of losing are mine. Perhaps a case can be made for an obligatory insurance against the burdens of involuntary accidents, but not for the risks we run voluntarily – starting with jumping from an airplane without a parachute and ending with taking a job in earthquake-prone California. I cannot oblige my neighbours to share the risks I take, not without giving them a say in my decisions on risks and thereby giving up my freedom to take risks. And once I give up my freedom to run one specific risk, say that of creating an accidental child, I give up any freedom to run any risk, in fact any freedom whatsoever – and incidentally take away anyone else's freedom as well.

Even a less radical liberal will have difficulty with the idea of an obligatory insurance scheme. It presumes that each person will or should appreciate risks in the same way (compare Rawls's maximin rule (1972: 152)) and this is a controversial view. Dworkin, certainly no libertarian, has argued that if liberalism is committed to at least tolerating moral pluralism, hence to the freedom of life, then we should leave it up to the individual to decide which risks they will run, which ones they will avoid *and* how well they want to be assured against bad luck – if at all (Dworkin 1981).

Three final remarks. First, the absence of a collective obligation to reserve resources for unplanned future generations does not imply that there is any reason why individuals should not insure their possible children against being created involuntarily. Nor does it in any way prohibit the creation of insurance schemes to pool the risks, nor even the administration of such schemes by political institutions – just as long as the scheme remains private and voluntary.

Second, the fact that liberal democratic societies are not obliged to put anything special aside for non-existing future generations in no way voids any of the rights that a voluntarily or involuntarily created child has as an existing member of an existing society. Its birth may mean a net loss in resources to others, but this cannot be blamed on the child or revenged on it. We will just have to learn to live with less.

Third and finally, there is one other thing we owe, metaphorically speaking, to non-existing future generations, something we actually owe to ourselves: the liberty to choose whether or not to have children. In this respect, distributive solutions can contribute greatly to social sustainability by improving sex education, by more evenly distributing information on family planning, by making it possible and making the means accessible, by stimulating the public debate about procreation so that procreation is no longer 'preferable', 'expected', considered instinctive or necessarily fulfilling, and childlessness no longer abnormal, shameful, regrettable – so that having children becomes an option and no more than that.

At this point, I want to rest my case. There is insufficient reason to believe that an obligatory insurance scheme can be justified as consistent with liberalism. It follows that there is also insufficient reason to believe that liberal democratic societies have a moral duty to invest in or save resources for involuntarily created future generations, beyond the level for which the restraint principle calls. I leave it to others to carry the discussion further.

7.5 Interspecies justice

Distributive approaches to sustainability presume the existence of subjects, the entities to whom rights are attributed, and objects, the things to which these rights relate. Up to this point, I have assumed that the subjects are humans and humans alone, whereas the rest of nature could only perform a role as objects of distribution. In this section, I add (some) animals to the set of subjects.

First let me recapitulate some points made earlier. We have seen that for purely technical reasons neither the non-human rest of nature nor any part of it can be a subject of rights. The language of formal rights does not allow this. Rights presume consciousness of choice and freedom of choice, not to mention other prerequisites that trees, rocks and animals simply do not meet. It is therefore only possible to attribute rights in a *metaphorical* sense to (parts of) non-human nature. In *formal* terms, these rights must be represented as freedoms, duties, that have either not been attributed at all or as duties and prohibitions attributed to humans. Both elements are present in the restraint principle: it demands that no part of nature be destroyed (an 'unattributed' right) unless necessary, and that in all other cases nothing be destroyed unless some sort of alternative is made available (a combination of duties and prohibitions).

We have also seen that next to humans, some animals deserve to be treated as subjects by virtue of properties like agency and consciousness. That is, in view of the previous point, they deserve to be treated *as if* they were subjects of rights. Like humans, they ought not to be treated merely like objects: in the attribution of rights, their interests should also be taken into account. In Kant's terminology, they should not be treated solely as means but also as ends in themselves. I also argued that the interests of animals are slightly more limited than those of humans in that the former do not usually have an interest in a continued existence.

Returning to the issue of distributive justice, it is now clear why a distinction should be made between humans and animals on the one hand and the rest of nature on the other. We do not owe the latter category any special treatment as subject(s), nor do we owe it special treatment as having intrinsic value (a notion we dismantled). All our (metaphorically speaking) obligations to nature are covered by the restraint principle. Whatever value this section of nature has is external, i.e. in the eye of the beholder. It is only in this latter respect, when a subject *j* values some object *x* as insubstitutable in an extremely strong sense, that *x* can be treated as special.

What remains to be established, then, is how to account for animal interests in terms of distributive justice. Since I already discussed the edge of their existence, that is, whether or not we have a principal right to kill them, I can limit the analysis to their lives, and since it would be absurd to see them as responsible persons who should or should not do certain things to us humans, it is possible to limit the problem still further: what do we owe them? I shall try to answer this question by discussing a few examples first. More general observations will have to wait until later.

Social sustainability is predicated on at least some form of 'purely' environmental sustainability. Hence, if the destruction of one single (but essential) element of an ecosystem would threaten the survival of the whole system, say, a swamp, and if this particular swamp were indispensable for social sustainability here and now or across generations, then society has an overriding interest in protecting rather than destroying the cornerstone of this ecosystem. This is an amendment to liberal democracy only in so far as it adds a new side-constraint: the same reasoning could be applied to more traditional prerequisites of a viable liberal democratic society like property rights, education and housing.

Let us next assume that the existence of our swamp is in no way a necessary condition for our survival but that it is for a particular species of animal living there, the swamp monkey. And let us consider two possible scenarios.

(1) If we were to turn the swamp into a forest, just because we would like to see a forest on that particular place and not because we need it, individual swamp monkeys would slowly starve and the species would become extinct.
(2) We do nothing whatsoever with the swamp, neither directly nor indirectly, but natural circumstances cause the swamp to turn into a forest. Individual swamp monkeys slowly starve, the species becomes extinct.

A recognition of the status of animals as morally relevant subjects does not commit us to preventing a species (whether human or animal) from becoming extinct: species as such have no interests, no feelings, no consciousness, no plans and, at least from a liberal point of view, it is only the life and welfare of individuals (with interests, feelings, etc.) that counts. Species are, like societies for humans, only important in so far as they sustain and promote the lives of individuals. Nor do the reasons why we should think of animals as subjects oblige us to prevent their individual deaths: we have no reason to assume that they have an interest in continued existence. However, since individual animals are conscious of harm, we are prevented from two other acts: from harming individual animals indirectly by killing their 'next of kin' with whom they cooperate and whom they need to survive, and from harming them physically. On the first scenario, we would be directly responsible for these acts since we are the ones who perform them. On the second scenario, it is the blind forces of nature that cause harm to the swamp monkey. Our responsibility here is indirect: we let it happen. If we cannot make a distinction between these two scenarios, the logic that obliges us to reject the first scenario will oblige us to reject the second as well. Yet if we were morally obliged to prevent the extinction through natural causes of the swamp monkey because we should not harm individual animals even indirectly, then we are obliged to prevent *any* harm from happening to animals. We would be obliged to provide animals with a system of social security, health care, shrinks, old age benefits, we would have to turn lions (being carnivores) into vultures or vegetarians, replace poisonous plants by non-toxic species. Although it may seem absurd, especially from the point of view of environmentalists who would want humankind to interfere less with nature, the inevitable conclusion of any argument that presumes that avoidable harm to animals *must* be avoided is that we ought to create a perfectly artificial world. Yet if we were to argue that not preventing avoidable harm is morally permissible, we would end up with a contradiction between this argument and our earlier conclusion that even (human) strangers in the desert have an obligation to keep each other from harm.

179

So why would omission be a sin when it concerns harm to humans but not when it concerns harm to animals? Because there is, or can be, a relevant similarity between the way nature operates and the way humans operate. Consider another scenario:

(3) We turn the swamp into a forest because we need the wood and because this is the only way to get rid of a species of lethal swamp insects. Individual swamp monkeys slowly starve, the species becomes extinct.

The similarity between scenarios 2 and 3 is that there is no 'natural' alternative that would save the swamp monkey. Nature cannot 'act' any differently, and neither can we. If we were to oblige ourselves to interfere in scenario 2 for reasons discussed earlier, we would also oblige ourselves to interfere in scenario 3: that is, to prevent ourselves from doing the only thing we can reasonably do and the one thing we are obliged to do – which is a logical contradiction. Since we cannot be obliged to do not-x when we are obliged to do x in scenario 3, we cannot be obliged to do not-x in scenario 2 either. In other words, we are not obliged to interfere when nature eradicates a species. We have to let go of the assumption that all harm is evil and that this obliges us to act against it.

It should be noted that although we are not obliged to interfere in scenario 2, we are not *prohibited* from interfering either. All we can establish is that liberal democratic societies do not have a duty to protect any species except itself against the forces of nature. It does not follow that we (as individuals) should no longer feed the ducks in the pond or the birds during a cold winter. More important, the absence of an obligation to interfere in nature implies that we are, in principle, free to develop a sustainable environment or let it develop itself in any possible direction we want.

If we do owe it to animals to take their interests into account, if, when and in so far as we are directly responsible for their welfare, then we owe them much the same treatment as we owe humans. That is to say, for one, that the restraint principle should be applied in their interest as well as in ours: no part of nature can be sacrificed to human desires if that would take away the means of existence for animals and no part of it should be sacrificed to human needs without offering animals a suitable alternative. Only in circumstances where human needs cannot be satisfied without sacrificing (the interests of) animals do the former prevail: humans have an interest in continued existence that animals do not (seem to) have.

Taking account of animals as subjects of rights almost like us also implies that we should treat them as objects of rights like us. Hence, animals have (metaphorically speaking) inalienable and unconditional rights like us, with the one difference that they do not have a right to a continued existence; their rights are valid only as long as they live or are left alive, ours are valid as long as we live. Focusing for a moment on the treatment they deserve while living, this implies, for example, that we may not just cut off a cow's leg for food and let the animal bleed to death or subject hens to life in a battery. Both these practices involve the undefended and indefensible violation of animals' unconditional rights to their basic needs, in this case the need not to be tortured or harmed; the first would also be a violation of the cow's inalienable right to the bodily parts it needs to exist. On the other hand, sheep as they exist today grow more hair than they need, cows produce more milk than their offspring needs (humans bred them that way). They do not therefore have an unconditional right to their excess hair and milk; we are free to use these for our own purposes. As a matter of fact, in these two cases we may even be obliged to use them since not fleecing or milking them could be harmful.

The catch is in the killing. The restraint principle tells us that if killing animals is necessary, they should be restored (perhaps cloned) or replaced by a similar animal, or that a proper form of compensation should be provided. Let us assume that nature can take care of this in the case of wild animals (and that we let it), and that we can either raise new farm animals to take the place of the old ones or that we allow nature to reconquer some environmental space.

More importantly, the restraint principle obliges us not to destroy 'objects of rights', in this case animals, unless necessary. If no alternatives to animals are available, the decision is a simple one, and if we face a choice between the lives of two animals (cf. the human life-or-death dilemma in Section 7.2), the decisive arguments will have to be found elsewhere. If there are alternatives, as when meat can be replaced by vegetables or pelts by textile, and given the fact that animals have no interest in continued existence, the decision on the necessity of killing depends on whether or not killing causes (permissible) harm. Now we do no more harm than nature does when we tear open a living animal and start gnawing away at its intestines while it slowly and painfully dies, but – parallel to the four scenarios described above – unlike nature we can have a choice in how we kill, thus be responsible for the pain we cause, and thus be obliged to kill as painlessly as possible. Yet unless we have reason to believe that killing 'as painlessly as possible' can be done without *any* pain, we are stuck with vegetarianism, except in exceptional cases. This is an empirical question and one that philosophy cannot decide.

The issue of the ethical treatment of animals is a very complex one, an issue moreover that does not always involve distributive or environmental questions or questions of sustainability. I have deliberately ignored themes like vivisection, zoos, cloning, genetic engineering, the existence of pet and farm animals, the creation of new species in general. It is impossible to treat all these topics at proper length in the context of this book. I only hope I have been able to show two things: first, that in so far as animal ethics is an issue of sustainability, it can be dealt with from a liberal point of view; and second that a green liberal position obliges us to treat animals with a certain degree of 'humanity'.

7.6 Dynamics and uncertainty

We have moved from a closed, one-generation, one-society state to a world in which multiple political entities operate, in which several generations coexist and in which the interests of animals are represented. Our model of distributive justice in liberal democratic societies would now be complete if it were not for the fact that there are two last variables which we have assumed to be constant throughout this chapter: population and resources. I represented these factors in the assumption that resources are only 'moderately', that is, economically but not environmentally, scarce. This meant that we assumed the existence of a constant stock of renewable and constantly renewed resources, a stock that would be large enough to warrant social sustainability and satisfy at least the basic needs of subjects, but not large enough to make questions of fair distribution redundant. I shall drop this assumption now and bring some dynamics into the model.

In modern societies, technology, science and the pursuit of wealth continually give us three new things: new technologies, new information on the advantages, risks and availability of resources and techniques, and new resources. Each of these three supply-side factors influence the size of a society's natural and artificial capital. Capital growth has a distributive and a supply-side aspect. If its growth is at least as large as population control, it can have a positive net effect on people's freedom: more goods will be available and – which is what really matters – more basic acts will be possible. If its growth is outmatched by population growth, it is by the same token likely to have a negative effect on freedom. These are pure supply-side effects, which I shall ignore here and discuss in the next chapter. Independent of these effects, changes in information, resources and technologies have distributive effects as well: they cause the amount and composition of the (rights to) goods available for attribution to change. It is on the latter aspect that I concentrate in this section. Incidentally, the notion of supply-side dynamics can also account for the effects of changes in population levels. Since (absolute) changes in population numbers can be read as new information on the (relative) availability of resources, I can ignore population effects as such and concentrate on the three factors mentioned, without pretending that all change in society really originates on the supply side.

At first sight, we would not or should not expect any fundamental effects of supply-side growth on distributive schemes. In principle, nothing changes if resources run out or become available on a wider scale. Both the general liberal principles of justice and the additional restraint principle will keep on doing their work as before; there is just more or less to distribute. Ideally, these principles will not only apply under circumstances of moderate scarcity but will also be good enough to offer moral (if perhaps not always practical) guidance under circumstances of extreme scarcity. Moreover, there is no reason to suspect that supply-side growth can ever lead to a Utopian state of abundance, i.e. a situation in which every possible desire can be physically satisfied. There is a mathematical point at which the total sum of desires equals the total amount of available resources, but there is no such thing as abundance, 'more than' enough. If all we desire are apples and we would have one apple more than we want, we have already passed the equilibrium point of perfect satisfaction and entered the stage where smelly rotten apples and the bugs they attract become a burden.

There is one hitch, though: all this assumes certainty. What supply-side growth adds is uncertainty. We cannot be blamed for attributing rights to a resource x at time t_1 in a way that turns out to be unjust at t_2 if we have unexpectedly either run out of x or doubled the stock by t_2. We can definitely be blamed if we knew for certain this was going to happen. But most of our real world decisions on the distribution of resources are made under conditions of uncertainty, or, to be more precise, under conditions where we have no knowledge of probabilities (pure uncertainty) and conditions where we can estimate probabilities (risk). It is at this point that politicians and green writers often invoke the so-called precautionary principle.

There is no avoiding uncertainty. Scientific knowledge is by its very nature uncertain and incomplete (albeit to varying degrees), developments in technology and resources are often unpredictable. Moreover, the beneficial and hazardous effects of supply-side growth tend not to be symmetrical. Growth always involves sacrifices: nature gets transformed into mining sites, factories and cities, funds that could have been used for direct relief are used for investments. And it is not always the beneficiary who pays. To deal with uncertainty, to decide whether we want to run risks at all and if so, how they should be distributed, we need a theory of risk evaluation. The precautionary principle represents one such theory.

To begin at the beginning: the only way to represent the relevant factors for policy choices under conditions of uncertainty is in the form of a cost-benefit analysis. The term should not be interpreted too narrowly, as representing only those quantifiable costs that can be expressed in money, but it nevertheless has its disadvantages. For one, whatever the currency used may be, we need *one* currency in which to express all costs and benefits, although this does not necessarily mean that, say, trees and computers are intersubstitutable. A cost-benefit analysis can also only measure total benefits and burdens, not their distribution over individual subjects. To account for distributive effects, we will need another meter. Finally, we have to be able to attach probabilities to costs and benefits, so that pure uncertainty can either not be represented or only imperfectly. There is one disadvantage that cost-benefit analyses do not have: they can (although they often do not) represent unacceptable costs or benefits, simply by expressing them as infinitely large.

In this case, we can choose to analyse only environmental costs and benefits or we can try to include them in a more general analysis – the difference is immaterial for the arguments I am about to make. The important things to remember are that if the benefits of a policy outweigh the costs, it is rational to consider it; that if the benefits of one policy exceed those of another, it is rational to prefer the first one; and that the 'green liberal' currency for costs and benefits is (freedom measured in) basic acts. Now consider how we assess risks. We do not just say that 'there is' a risk and that 'therefore' something should be done; instead, we put a weight on risks: some are worth taking, others are not. And we assess the value of whatever is at risk: a 90 per cent chance of one single ant getting killed is usually considered far less important than a 20 per cent chance of a 10 feet rise of the sea level. At first sight, we would say that we evaluate the cost side, the environmental risk side, of a policy as follows:

$$a \times (\text{probability})^b \times c \times (\text{costs of possible effect})^d$$

in which the weighing factors a and b depend on one another, as do c and d.

We have, according to this formula, one standard to evaluate the costs of, say, pollution and another, independent from the former, to evaluate risks as such. Hence, if we put a high (dis)value on risks, we have more reason to act – or in other words: the greater the uncertainty, the greater the need for precaution. Environmentalists often refer to this idea as the precautionary principle, and oppose it to what they see as the traditional attitude of policy-makers: to interpret uncertainty not as risk but as the absence of contradictory information. Over the last ten years, particularly since the 1992 United Nations Conference on Sustainable Development in Rio de Janeiro, the precautionary principle has also been introduced in 'modern' environmental policy. Like nature, the precautionary principle is a concept that can be and is interpreted in countless ways. Sometimes it is taken to mean that the existence of potentially serious risks is enough to justify precautionary action (i.e. preventing risks), sometimes serious risks are necessary but not sufficient to justify precautionary action, sometimes risks have to be weighted against costs, sometimes risks and costs of a non-environmental nature are included and sometimes they are not (cf. Francis 1996) – and there is little agreement among the wise on when a risk is 'serious' or 'considerable' (cf. Jordan and O'Riordan 1995).[36] Despite these permutations, there appears to be a series of core ideas shared by all conceptions of the precautionary principle, for example a propensity for action rather than inactivity, for shifting the onus of proof, for consideration of medium- and long-term effects, for a certain proportionality between the costs and the effectiveness of precautionary action (cf. Bosselmann 1992; O'Riordan & Jordan 1995). In short, the overruling feeling seems to be that the existence of uncertainty (here: about the environmental effects of human actions) is enough to put the onus of proof for the acceptability of potentially environmentally dangerous activities on those who propose such activities.

That the precautionary principle is a sound principle is a widely shared view but, I think, one that is basically incorrect – as incorrect, in fact, as the so-called 'traditional' conception of environmental risks. It might seem an academic point but I hope the reader bears with me. First, consider this: the weight of risks themselves seems to depend on the value we attach to the object of the risk. Imagine that all you have to do to protect yourself against pickpockets is to wear your wallet upside down; the costs of doing this, the efforts involved, are really negligible. Now a 50 per cent chance of losing ten pence usually does not incite us to take the same or more precautions than a 50 per cent chance of losing £500,000, or a 25 per cent chance, or even a 5 per cent or 1 per cent chance. We simply do not find the risk of losing ten pence worth considering – because we care less about the money, and because we care less about risks *because* we care less about the money. Hence, risks are evaluated by one and only one standard: the value we attach to possible effects:

$$c \times (\text{probability})^b \times (\text{costs of possible effect})^d$$

in which b and d are both functions of c.

Now note the consequences of this view. In ordinary discourse, environmental probabilities are used in two ways: proponents of precautionary environmental policies argue that there are risks and that something should be done about them because they are risks, whereas opponents argue that they are uncertainties and that we should at least wait until we know more (cf. Weale 1992). The effect of my interpretation of risks is that probabilities themselves count for nothing; they have no independent role. (I shall amend this idea in a moment.) Hence, any appeal to the concepts of risk and uncertainty to either do something or do nothing are misleading and at times simply rhetorical. What matters when we evaluate environmental policies is not the fact that we are certain or uncertain about effects, but how we weigh those effects, how we value, say, a polluted and unpolluted beach.

What the precautionary principle does is obscure a fact that serves as a central axiom in liberal thought. It conceals the fact that we disagree on the value of parts of our environment, on their value relative to other things than the environment (some of which the precautionary principle would want us to sacrifice) and, in general, on the good life. The principle thereby also misrepresents the role of risks and uncertainty in our lives. Our attitude towards risks is a part of our individual conceptions of the good life, not something that exists independently. Some of us are gamblers, others seek certainty; some want to minimize their chances of being worst off in all or some situations, others hope to maximize their chances of being best off. It may be true that people can make irrational *estimates* of their chances, and even that a race like the system gamblers in casinos are ineradicable no matter how much statistics they are taught, but an irrational estimate of our chances is not the same thing as a different appreciation of risk as such (cf. Section 5.3).

The precautionary principle is not an acceptable theory of risks for a liberal democratic society. It is a principle that only individuals can use as a guide for individual or perhaps cooperative action, since it presumes something that is not likely to be found at a macrolevel: consensus on the goals and criteria of social sustainability, on the good life, on theories of the good in general and on risk appraisal in particular. Yet policy realms need some kind of uncertainty principle in order to be able to answer the 'What if?' questions that are bound to be raised by (and against) any particular distribution of rights over a society. We know which criteria such a superior principle must meet: it must be able to deal with both forms of uncertainty, risk as well as pure uncertainty, and it must be consistent with the liberal appreciation of moral pluralism including diverging opinions on the value of uncertainty.

Since the intention of this book is to establish the possibility, and not to proclaim the authoritative expression, of green liberalism, I will not try to hand down from high up the one and only true form of this uncertainty principle. However, there is one slightly more modest claim I would like to make: whatever form be given to a liberal uncertainty principle, it will necessarily presume or even contain the restraint principle. We have already seen (in Section 6.1) that the restraint principle is neutral with regard to individual attitudes towards uncertainty: both Rawlsian risk-avoiders and non-Rawlsian gamblers would accept it as rational. What we should also realize is that it is permissive with regard to individuals' appreciations of uncertainty and at the same time quite strict on how the costs of voluntary bad luck should be distributed: anyway, as long as it is not at other subjects' expense. Finally, it distinguishes between risks that must be taken ('necessary' risks, risks made in the pursuit of a full life, of basic needs) and those that are only conditionally permissible.

Environmentalists should note that the restraint principle and the precautionary principle are identical in one key aspect: they put the onus of proof on the side of those who create the risks. They may see an extra advantage in the restraint principle (relative to the precautionary principle) because the former can also account for pure uncertainty, whereas the latter cannot.

In this chapter, I considered the effects of the introduction of environmental criteria, in particular the restraint principle, on liberal theories of distributive justice. I have been talking about principles and the limits they pose on distributive policies, not about real, solid, concrete, tangible distributive solutions for the problem of living within the limits to growth. The reader may by this time wonder what actually happened to that part of the story. Where is the international redistribution of power, technology and natural resources for which so many green parties and groups plead so intensely? Who will do what about the deterioration of the ozone layer, about the greenhouse effect, rising sea levels, export to Third World countries of First World pollution, the depletion of natural resources, the future of our great-grandchildren?

This call to go back to basics is understandable but – I hope – inappropriate. In so far as this is not a contradiction in terms, I have been implicitly concrete. I have argued, among other things, for a rigorous analysis of the environmental effects of any policy that involves the distribution of physical goods. I have argued that the production of waste and pollution imperils and sometimes violates the rights of individuals, that waste should be treated not as the leftovers of yesterday's feast but as an integral component of resource management. I have argued that there are strong obligations in the field of international justice, and I have reaffirmed that individuals should not (be able to) duck their responsibility for the children they create. I have defended the radical position that the lives and ecosystems of animals have a value independent of the value we humans attach to both, so that human wants cannot all that easily overrule the needs of animals. I have argued for principles that are at least as concrete as 'the polluter pays' and 'the precautionary principle' – the alternatives may be aesthetically inferior but they are vastly superior from an ethical point of view. I could not have been more concrete without abandoning the theoretical perspective in favour of empirical facts and figures, and this is precisely what I wanted to avoid. Facts are interesting and important. Whether or not, say, a redistribution of cars resulting in a globally equal distribution of the freedom of movement would be environmentally good or bad is extremely relevant to the question of social sustainability, as are the effects that a more equal distribution of different types of pollution may have. Yet these are matters that have been dealt with by others far more accurately than I can ever hope to do. While I admit that distributive policies can have positive as well as negative effects on social sustainability, what I have tried to do is not to ask *which* precise effects which precise policy would have, but to see *how* we should assess distributive policies and their effects. We need guiding theories that answer the deeper question why we actually should do whatever we can do, before it can make sense to do anything.

8

Supply-side politics

8.1 Tactics

Of the three stategies for sustainability, I have now discussed two – with rather disappointing results for sustainability. Population policies cannot guarantee social sustainability even under the unlikely assumption that we have full information on when exactly the upper limit of a sustainable population volume would be reached. In a liberal democratic society, population control must depend on voluntary co-operation. Distributive solutions to problems of sustainability can only ameliorate a problem by distributing the burdens, not solve it: they determine *who* uses nature to what purpose, not *how much* of it is used. If we really want liberal democratic institutions to safeguard sustainability, the only option left now is the application of supply-side policies. In this chapter I argue that supply-side policies can improve our chances considerably but still not guarantee a definitive solution.

Basically, the idea of a supply-side answer to questions of sustainability is simply to increase the amount of available resources within a given policy realm. I assume that a policy realm can be equated with 'a given and constant demand'. In other words, I assume either a constant population size and constant size, weight, appetite and diet of all individuals, or an in sum constant but individually variable size, weight, appetite and diet of a variable number of individuals. I also assume that 'demand' explicitly covers the demands of non-human subjects and implicitly covers all necessary conditions for the continued existence of the subjects' environment (as being in the interest of subjects). These assumptions must be made to avoid further and unnecessary complications. If there is a reason why supply-side solutions necessarily cannot guarantee sustainability given a constant demand (which is what I shall argue), they surely cannot guarantee sustainability under unstable circumstances.

The amount of available resources can be increased in many ways, although these are all variations on two themes: adding new resources and using existing resources more efficiently. I distinguish the following types of supply-side solution:

(1) External growth of the resources of policy realms by conquering space
(2) External growth of the resources of a policy realm by annexation of new territories on Earth
(3) External growth of the resources of two or more policy realms by cooperation
(4) Internal growth by discovering new stocks
(5) Internal growth by discovering alternative resources
(6) Internal growth by using existing resources more efficiently, given unchanging technologies (where efficiency equals minimal damage to the environment)
(7) Internal growth by using more efficient technology[37]
(8) Intensive growth by using consumer goods more efficiently
(9) Intensive growth by means of recycling.

Not all these techniques need to be discussed in detail. An expansion into space can be disregarded for practical reasons. It is generally considered to be infeasible at this moment in time,[38] if it were feasible it would still be inefficient, and if it could be efficient it might still not be enough to solve all our supply-side problems. Assuming that it could be all of this – a feasible and efficient panacea – expansion into outer space could offer a solution for many generations to come, even if the demand increases. Yet – this is presently still an academic point – unless human expansion can reach as far as the universe itself, i.e. into infinity, expansion will always remain a temporary solution: as long as there are limits to expansion, there are limits to the amount of resources available and therefore to the size of the demand that can be satisfied. (For an introduction to the relevancy of space exploration and its problems see Hartmann 1984.)

There may also be other considerations. Ecologists could, for example, add that whatever practical advantages expansion into space can have, they would only carry the human way of life to the stars and change nothing about the human attitude of selfish, short-sighted, ruinous exploitation of, and total disrespect for, nature. This is an irrelevant objection: in so far as our present topic concerns the possibility and not the permissibility of sustainability policies, ethical arguments are simply inappropriate; and in so far as it is an issue of permissibility, expansion would have to conform with liberal democratic principles including the restraint principle. The ecologist's critique would then have to be directed against this latter principle and not against its application to a particular case.

It is therefore for purely practical reasons that I shall ignore the possibility of solving earthly problems by hopping planets. Whatever contributions expansion into space can make nowadays, even in combination with other techniques, will be too small to offer immediate relief or a long-term solution. Moreover, these hypothetical contributions do not differ too fundamentally from what I have called 'internal growth' techniques: the discovery of new stock, of alternative resources, of new technology and of ways of using existing technology more efficiently.

A second technique that can be disregarded is that of the annexation of new territories. Annexation is even more than expansion a temporary solution; it simply moves the problem of limited resources into the future. Besides, it is only a partial solution, creating at the very best 'sustainability in one country'. If annexation involves conquering other subjects (i.e. both humans and animals), the conqueror moreover becomes morally responsible for the attribution of rights to the conquered (cf. Section 7.3), and by conquering they arbitrarily limit those rights – arbitrarily, because there is no impartial reason why subjects in one territory deserve to suffer involuntarily for the benefit of those in another. If annexation does not involve conquering other subjects and if there were no principled objections against annexation (the loss of liberty, for instance), it can still be unjust, at the very least if not 'enough and as good' is left for others: again, there is no impartial reason why the options of one policy realm should be sacrificed for those of another. Finally, in both these last two cases it is imaginable that no one is or needs to be harmed as far as natural resources are concerned. In that respect, the difference between annexation and cooperation (the third technique) lies in the fact that the conquering party gets all the benefits but does not deserve them any more than any other excluded party.

Even cooperation, which is basically a redistributive and not a supply-side solution, begs to be ignored: in so far as it *is* a supply-side solution, i.e. an increase in resources for distinct policy realms, it differs in no way from internal growth by discovering new or alternative resources.

Before we consider the liberal position on the remaining supply-side techniques, I want to describe briefly how greens – environmentalists and ecologists – view them. Since they tend not to be neutral but to distrust many of these techniques, it would help to know why. We have already encountered the two most conspicuous green arguments against supply-side solutions. They do not change the (Western) human way of life, which is seen as a major cause of the environmental crisis, but instead prolong it. I replied to this argument above by saying that in our context, this should be read as a possible critique of the restraint principle rather than as a critique of specific solutions. Liberalism, once amended with the restraint principle, goes as far as we can legitimately go in dictating limits to the satisfaction of individuals' preferences. If it does not go far enough, political decision-makers need either different and more severe political limits or they will be forced to interfere in the preference formation process itself, i.e. in 'the' human way of life. As I argue in more detail in Chapter 9, both methods lead to the abandonment of liberal democracy in favour of far less attractive political systems.

Furthermore, supply-side solutions often do little more than move a practical problem forward in time. The discovery of new oil fields, the invention of more economic machinery or of mining techniques that allow mining at greater depth may all add resources but never enough to ensure that *any* demand can be met: the Earth is finite and infinity as yet unconquerable. All this is not to say that we should dismiss supply-side solutions forthwith as totally irrelevant, but, greens argue, we should be mindful of their practical limitations. They can only offer temporary relief.

The problem with this argument is that it loses its relevancy somewhere along the way. It hinges on empirical, read contingent, facts and possibly also on a questionable interpretation of 'limits' to the availability of resources. It can certainly be correct when directed against any one particular supply-side solution but whether it will also apply to mixed solutions is doubtful. We should first bear in mind that all resources are available in limited quantities, i.e. they are 'depletable', but some are more depletable than others. It seems reasonable to assume that our sun (our source of solar energy) will go on shining for millions of years but one day it will burn out. As long as the sun shines, trees can infinitely be replaced within ten to a thousand years, yet wood itself is a more depletable resource than sunlight: unless individual trees are replaced on time, our stocks will decrease and ultimately vanish. What matters then, in regard to the depletability of resources and the quest for sustainability, is not that all resources are depletable (thus making perpetual sustainability mathematically impossible) but the *rate* at which resources can be renewed. In other words, sustainability must be either logical nonsense or an attempt to prolong the quest for survival as long as *possible* and not for ever. Learning to live with limits to growth therefore does not just mean that our demands must not be larger than the available amount of resources but also, first and foremost, that there must be a kind of *balance* between the use and renewal of resources. Note the 'kind of' in the last sentence: since all resources are ultimately depletable, a perfect and permanent balance is a fiction. Iron, for instance, is a far from renewable resource – far more depletable than wood or solar energy. Hence, given our present reserves and demand, we will run out of iron one day. More economic use of iron ore and the discovery of new supplies will postpone that day.

So far, the practical objection that supply-side techniques only postpone the inevitable is perfectly correct. The invention of new or more efficient mining and production techniques can prolong the quest for survival even longer, and the objection still applies. Iron can be regained: we could try to recycle old goods and thus use – theoretically – every atom of iron used before time and time again. Next to this, we can improve the efficiency with which consumer goods containing iron are used, i.e. reuse rather than recycle old goods and prevent them from lying around unused. As all this still leaves us with an empirically limited stock, the objection still applies. Iron can also be replaced – by other metals, by organic or artificial materials both (as yet un)discovered and (as yet un)invented. The objection still applies. Ultimately, we may invent totally new consumer goods that replace the objects for which we use or used iron. The objection still applies, but by this time no longer to iron but to the new material. And by this time, the objection has also crossed the line between the sane and the absurd. There is no reason to assume we can go on inventing new and more economic goods and techniques indefinitely, but since perpetual survival is not to be expected we do not *need* to economize infinitely either. All we can hope for is to find a means of sustaining life for as long as possible, and this is exactly what mixes of supply-side solutions can do. To cut a long story short, the objection that supply-side techniques only move our problems forward in time is valid and practically helpful as an objection against particular techniques but it fails as an objection against supply-side techniques as such.

A further green objection to supply-side solutions can be voiced independent from the previous couple: they do not change the fact that material needs are infinite – where 'need' is defined as whatever is necessary to function without disorders (cf. Hilhorst 1994: 262). Let us assume that there are no moral objections to 'the human way of life' and that we employ all possible supply-side techniques on all fronts to satisfy – for the sake of argument only human – needs. It is then still possible that our needs are greater, and even grow at a higher rate, than our potential for satisfying them. An example (we could choose any technological gadget humans have ever invented) is the introduction of the personal computer: it has on the one hand improved old ways of satisfying needs (in Hilhorst's sense) by replacing typewriters, calculators, books, etc., but it has also created new needs: few individuals and even fewer companies can afford to lag behind on those who use the most advanced technology, hence nearly everyone is more or less forced to replace typewriters by PCs, to replace old monochrome screens by ever better colour screens, to attach soundblasters and superior printers, to create virtual networks, to link up with Internet, to enter the World Wide Web.

I admit this is an interesting objection, one that can be modified to apply to many other and possibly all sensible interpretations of 'need'. Yet its relevancy is again limited, since it seems to confuse two types of growth: growth in the number of needs and growth in the volume of the resources used. The introduction of new technology may lead to the development of new needs, but the objection gives us no reason to assume this will always be the case; the amount of resources required to satisfy these new needs may well more than offset the economizing effects of replacing old technology by new. But the objection again fails to show that this must necessarily follow; and finally, it fails to prove or even make it plausible that the amount of resources needed to satisfy the sum of old and new needs by means of new technology must be larger than the amount of resources required by old technology. All in all, our third objection is more a reminder of the fact that supply-side solutions do not have to be effective than a refutation of each or any particular technique.

Finally, greens of many denominations distrust 'technology'; some even reject it completely. The reasons are complex and vary with the interpretation given to the term technology. On some versions, technology is very broadly interpreted as a way of thinking (a refusal to accept nature as given) plus all that it results in. A slightly more modest version associates technology with Renaissance and Enlightenment ideas: it starts with the typically human refusal to accept nature as it has been given, develops into the belief that the laws of nature can be understood, that the universe is malleable and that humankind can be its demiurge, and it ends with the invention of (material) instruments and (psychological) mechanisms to transform the world – with the introduction of genetic engineering, even to create the new human. A third, shallow interpretation is that of technology as instruments and mechanisms only, a fourth even reduces it further to applied science only. A fifth distinguishes between technology in a strict (shallow) sense and the necessary social preconditions that determine the precise form of technology as a historical phenomenon (Van der Wal 1994). Each of these interpretations gives rise to a different critique of supply-side techniques.

The more shallow critics of technology point to practical obstacles. Technology often includes attempts at increasing efficiency and may overdo things, resulting in overfertilization, eutrophication, soil degradation, desertification. If technology aims at economic efficiency only, it may even lead to the development of instruments and goods that break down faster than would be technically necessary (what Schumpeter called 'creative destruction'). We can ignore critique of this kind here – it is obviously correct but (*ceterum censeo*) insufficient to be able to dismiss supply-side solutions as by definition counterproductive.

The more fundamental critique of the protechnology 'mind-set' is far more interesting. It takes roughly two forms (although the two are often confused; see e.g. Van der Wal 1994): a naturalistic and a psychological form. On both perspectives, technological solutions to environmental problems will be useless as long as humankind attempts to 'strive against the stream' (Van der Wal 1994: 243), i.e. reshape nature into something it did not and could not evolve into on its own. One way to make sense of this form of the argument is by assuming that the laws of nature do not allow us to rearrange nature to our liking – it will try to restore itself to what it was 'meant' to be, or it will malfunction and ultimately crash. If this is to be read as an appeal to abandon our evil ways and live in some kind of pastoral state of harmony with nature, the argument backfires: the refusal to accept nature as it is given is part of human nature; abandoning it is recreating nature to something it was not 'meant' to be. If it is a warning against transgressing the boundaries set by the laws of nature, it turns into the psychological version of the argument: it does not reject the human urge to adapt nature as 'unnatural' nor deny that nature is malleable but points out that natural laws themselves are not malleable, i.e. that there are limits to the kinds of world technologies can create. Of all the objections discussed in this section, this is the only one that makes sense as an objection against the *principle* of seeking supply-side solutions for social sustainability. It is the only one that allows us to say – or fear – that there is a point of no return in human history, a point past which ecological problems or even a total crisis become irreversible and irreparable.

No matter what we think of green critiques of supply-side solutions, it will by now be understandable both that and why greens prefer some supply-side solutions (recycling, increased efficiency and the use of renewable, if possible indepletable resources) over others, and why it has taken a long time before some of them could admit that 'technological' solutions (washing machines, central heating) can sometimes be more environmentally benign than 'natural' solutions or nature itself (Dobson 1995: 98; De Geus 1996: 184). Supply-side solutions, even highly technological ones, can be compatible with a green attitude. The question now is: how should green liberals assess supply-side solutions?

8.2 Green technology

Supply-side solutions aim at increasing the amount of available resources relative to the volume of the demand, i.e. the size and needs of a population. We have assumed the existence of a fixed demand, but even if demand and supply are balanced and there is no immediate danger to social sustainability, supply-side techniques cannot be or be made superfluous. As I remarked in the previous section, all natural resources are depletable – some faster than others, some with and some without human intervention, yet ultimately they are all finite. Hence, the aim of sustainability cannot be perpetual sustainability but merely prolongation of our (or some) existence. Policies aimed at sustainability should therefore be aimed not only at balancing demand and supply but also and foremost at maintaining as much of a balance as possible between the depletion and the renewal of natural resources. Given a constant demand volume, sustainability then requires a 100 per cent natural renewal degree which, since especially anorganic resources are hardly renewable, seems to much too ask. For this reason even an ecologically balanced society needs supply-side solutions – solutions in plural, for resources cannot in practice be fully recovered by recycling. A society like this may well need more than two supply-side techniques: more efficient utilization of old goods alone (for example) may not be enough to fill the gap left by recycling. The more realism we allow for the parameters of our problem (decreasing resources, increasing population), the more evident it becomes that supply-side techniques are indispensable.

Up to a point, the problem of social sustainability is still a purely academic problem. We, humankind, have not by far reached the physical limits to the exploitation of nature yet. We have, for instance, the technology to replace non- and less-renewable sources of energy by solar, wind and water power, we can replace slow-growing trees by faster ones, we have artificial fertilizers to improve food production and even artificial soil to improve on the natural stuff, we have hardly explored the oceans and are only just discovering the enormous potential of biotechnology and genetic engineering. In other words, we have probably not yet reached the level at which our demands equal our resources. Nevertheless, even under these fortuitous circumstances there is reason to use supply-side techniques for other purposes than the satisfaction of human desires; more precisely, there are (at least) three good reasons for using them to promote sustainability.

In the first place, even if we can go to the absolute limit of theoretically available resources and theoretically feasible supply-side solutions, we still have to find a balance between the depletion and the renewal of resources long before that. At any given moment, we have to make do (roughly) with the resources that are available at that time and no other, since new resources and techniques are not around yet. It is certainly reasonable to expect new discoveries and inventions, but there are limits to what we can reasonably assume them to be able to add to our resources and to how soon they can do that. Hence, the practical limits are more stringent than the theoretical limits. For liberals we might add that if we were to 'overgraze' on resources, we would use them without replacing them or allowing them to be renewed, and unless we have no other choice, this implies that we violate the restraint principle. Resources are means, means are options, options are possible rights and possible freedom. The duty to prevent this kind of overgrazing implies that every permissible supply-side technique should be used to that end.

Next to this, resources are not perfectly substitutable except under very special circumstances. An excess amount of plutonium cannot be set off against a shortage of grain unless we are plutonium- or rice-eaters, nor can excessive production here compensate for excessive pollution elsewhere unless 'we' value production and 'they' could do with a bit of poison in their environment. In short, the need to find a balance between depletion and renewal applies more to each type of resource separately than to the lot of them together. Hence, the practical limits are *far* more stringent than the theoretical ones.

Finally, as claimed in the previous section, the laws of nature put limits to what we can do with this world. Some imaginable worlds are practically impossible, some worlds that once were possible (e.g. one where we find a dodo in every zoo) have become impossible, worlds that are possible today will be impossible tomorrow. In general, each time we use a resource we limit the number of possible future worlds, and even though there is always a near-infinite number of possible future worlds, we may at any time exclude (a number of) those that are desirable, sustainable or both. If the exploitation of resources is not balanced against their renewal, even if we have not yet reached the limits of our existing resources, we exclude options both for future worlds and for present persons. For a liberal, this may not be enough to justify more stringent restrictions on the use of resources than the restraint principle already imposes, but their commitment to liberty and especially the liberty of life gives them an interest in promoting all permissible supply-side techniques as a means of extending liberty and prolonging the existence of one of its necessary conditions: a sustainable society.

There is no need to ask, then, if liberals should in principle approve of supply-side solutions; we have just seen that they should and why they should. Nor do we have to dwell too long on the permissibility of each of the nine techniques described in the previous section. Unlike some greens, liberals can accept them all without hesitation. Each and every one of them is compatible with our stringent conditions of liberal democracy, many of them might even improve the liberty and proportional equality of humans. Moreover, each technique is compatible with the restraint principle. One of the functions of supply-side techniques is to enable us to exploit resources to the maximum, i.e. to use each resource as efficiently and economically as possible, which links up perfectly with the demand of the restraint principle that resources themselves are managed as economically as possible.

Yet we should make a distinction between liberals' principal approval for supply-side techniques and the obligations of liberal democratic polities with regard to their employment. Approving does not imply encouraging. Liberal democratic political institutions are abstract entities; they cannot create or discover resources or invent new technologies. All they can do is attribute rights, and in this way they can stimulate research and development into supply-side solutions and facilitate the implementation of better technologies. Whether they should, and if so, how, is open to debate. So far, I have stressed only the liberating side of supply-side solutions, but for once I agree with a piece of green rhetoric: there is no such thing as a free lunch. Supply-side solutions come at a price. New resources frequently mean more pollution, more destruction, less nature; new technology often eradicates old ways of life (the short, nasty and brutish life in a typing pool) and introduces new and addictive desires. On the whole, supply-side technologies can but do not *necessarily* enlarge liberty, improve equality or make a sustainable society greener (cf. Hilhorst 1994: 268).

It seems that this objection is partly inappropriate in our context. The distinctions made earlier between conditional and unconditional rights and between weak, strong and absurdly strong substitutability allow us to say that if a supply-side technique would destroy anything that is insubstitutable – a forest, a species or a way of life – and thus the object of unconditional rights, then applying the technique cannot be permissible. Moreover, many of the options that (could) vanish and rights that are (could be) voided by new technology or resources disappear for justifiable reasons: with the disappearance of e.g. the open hearth in favour of a rediscovered classic technique, central heating, we witnessed or in most countries will witness the disappearance of an inefficient, environmentally inferior heating technique that, by using far more resources than necessary and producing far more waste than necessary, now needlessly infringes on other people's life and liberty. In short, if governments are there to attribute and protect rights, they do have a role in introducing new supply-side solutions. Yet there is still more than a grain of truth in the objection that new technology or the discovery of new resources does not have to be propitious only. Although I shall make some amendments later, the tragedy of the quest for sustainability is that it leaves liberal democratic decision-makers no choice as regards the promotion of research and development. They are ordinarily obliged to encourage technological solutions to environmental problems. I call this the paradox of technology. Unlike the rapid reproducers paradox, this one is not about doing the right thing and punishing the wrong people, this is the exact opposite: doing the 'wrong' thing (i.e. technology) and rewarding the right people.

Sustainability demands on the one hand that even though humankind has not reached the theoretical limits of its resources yet, it should still try to find a balance between depletion and renewal so as to protect the physical basis of its existence against destruction. On the other hand, sustainability is social sustainability: it requires that humankind be enabled to perpetuate its existence under tolerable circumstances. There is a conflict inherent to these two aims of sustainability. I have described 'tolerable circumstances' in terms of the criteria of liberal democracy and 'a life worth living', but the conflict in question does not depend on the level of luxury we live in; it will surface regardless of the exact phrasing. The second aim implies that political institutions should seek to create the circumstances under which people's lives can be at least tolerable. In a liberal democratic society, that means – among other things – protecting liberty and offering choices. Since humans need resources and since many resources are both finite and non-renewable, there exists a need for the discovery of new (stocks and types of) resources and new technology. Yet every bit of new technology and every bit of newly acquired resources represents an intrusion on nature: either we turn more of nature into natural capital or we squeeze more natural capital out of the same piece of nature. The more nature is capitalized, the more balances there are to keep, the higher the risk of instability becomes and hence the more we endanger the physical basis of our existence – despite the fact that the result could be a greener world. Seeking for a life under tolerable circumstances, the second aim of sustainability, thus leads to endangering the first aim of sustainability.

The tragedy in all this is that there is no alternative. The two imaginable alternatives are both Utopian and offer respectively an unacceptable alternative and no real alternative. Going back to nature, offering humans only one way of life based on perfectly renewable, nearly infinite resources under the condition that they do not in any way overgraze on their environment, is unacceptable. First, it is not a tolerable life for most present humans and it would require rigorous totalitarian de-education on a Utopian scale to adapt the tastes and preferences of future humans to a life in blissful primitivity. Second, it is naïve to expect that history will not repeat itself in some way: human beings tend to be interested in finding ways to make life easier or improve their position relative to others – both of which are incentives for reinventing technology. Third, from a liberal point of view a back-to-nature policy rewards people who do not need to be rewarded (very radical deep-green non-materialists) and punishes all those others who deserve no punishment by imposing one lifestyle on all. The second alternative is a great leap forwards to nature, i.e. discovering ways of living under back-to-nature rules while still having the freedom of life and lifestyle of a tolerable life. Even if it were imaginable, this cannot really be an alternative. We cannot live like this yet, so if it is possible at all, it requires infinitely more research and technology than we have nowadays to get there in the first place.

Hence, there is no escape from supply-side solutions. The same reasons that justified the psychological transformation of nature into resources where distributive solutions were concerned (cf. Section 7.2) justify research into and development of new resources and techniques. People (and animals) matter, they need resources, therefore it is legitimate to search for new resources and develop appropriate techniques. None of this invalidates green criticisms of technology in the broad sense. There is still every reason to be wary of the less healthy effects that technology can have: the creation of new 'needs', the increased dependency on what are supposed to be merely tools that make life easier, the inclination to confuse short-term patchwork with long-term solutions, the desire to see the malleability of nature and ignore the inflexibility of the laws of nature, and most of all the 'technocratic' attitude, the belief that there is a technological solution for every empirical problem including that of finite and depletable resources. We have to be wary of these risks, but we should also be careful not to confuse contingencies and principles. Without supply-side techniques, social sustainability is impossible; with them, it may just be possible.

8.3 Biodiversity and policy diversity

In this and the preceding two chapters we have looked at the three faces of the politics of sustainability. We excluded the possibility of 'demand control' in general and population politics in particular as incompatible with the stringent criteria of liberal democracy. Distributive approaches to ecological issues, on the other hand, were judged to be necessary and desirable from the point of view of a liberal who accepts (as I argued they should) the restraint principle. Yet although (re)distributive measures may be just, we also found that they cannot ensure sustainability. At best, a just attribution of rights can ameliorate environmental problems. Finally, I briefly discussed supply-side solutions and argued that liberal democratic political institutions have a duty to promote research directed at this type of solution, as well as a duty to promote the use of supply-side techniques provided they are compatible with the restraint principle and our stringent conditions of liberal democracy. All in all, the results may seem rather disappointing. We found the possibilities for sustainability to be limited in several ways: because all resources are depletable, because the supply-side factor (i.e. the discovery of resources and the development of new techniques) is unpredictable, because the demand-side factor is uncontrollable and because distributive solutions can only offer temporary relief.

However, things are not as bad as they look. We have achieved two goals. We have established that it is permitted for liberal democracies to develop policies aimed at improving the chances for social sustainability. In doing so we have also made a theoretical point: we have taken position on several of the green issues mentioned in Chapter 2, thus showing that liberalism can at least *discuss* environmental questions. Whether this means that liberalism can be green enough to make 'green liberalism' possible depends on more factors than political positions alone; we shall reserve judgement on that until the last chapter. Before I turn to the principal topic of this section, liberal policies aimed at sustainability, I should first like to list the theoretical choices made on the green issues of Chapter 2 – just for the record. I did not take position on the dimension of the scale of environmental problems since this is an empirical matter, nor have I said anything about whether or not liberals actually and/or currently believe in limits to growth. I simply assumed a finite world and showed that, given this assumption, it is rational (for liberals, greens and everyone else) to accept the existence of limits to resources, of limits to all solution strategies, and of a need to balance the renewal and depletion of resources. Hence, even though the quantitative development of society (roughly: population growth) cannot legitimately be controlled, material growth (resources and techniques) is acceptable only within limits: liberals prefer zero growth but do not exclude further growth if necessity demands it. I have not yet discussed the issue of qualitative (economic) growth; I shall do that in the next chapter. Finally, although we assumed that in a liberal account of environmental politics the well-being of society must matter, we also saw that its importance is limited. Since animals should be regarded as moral subjects and since nature is the physical foundation of our existence, the fate of human society, however important to liberals, cannot be disconnected from that of the rest of nature.

Let us turn to policies. In Chapter 2, I suggested that liberalism is compatible with reformist policies only, preferably at a collective level. Political action at the individual and group level would be permissible only if it were compatible with 'real' policy, collective action. The last three chapters have more or less confirmed this supposition. Liberals prefer a low profile, that is: the least socially disruptive policy that empirical circumstances allow, and the best possible policy as regards the just distribution of the liberty of life. The existence of certain criteria for a liberal democracy imply that liberals should permit environmentally friendly or hostile action by individuals and groups within society only within the limits set by these conditions, by the restraint principle and by the necessity to show proper respect for individuals' conditional and unconditional rights.

This is all very abstract. The reader would probably – and I would certainly – prefer it if it were possible to describe in exact and full detail which environmental policies liberal democracies should implement and which ones would not be permitted. However – and this will be the core argument of the present section – such precision is neither possible nor desirable. First, there are practical reasons why environmental policies cannot be the same all over the world so that nearly any policy x (out of the infinitely many that we can imagine) can be totally appropriate in one place and totally inappropriate in another. Second, policies are not morally neutral. They use instruments that can or cannot be permissible depending on circumstances, and they are designed to create a more desirable world – but liberals have a very peculiar conception of what a desirable world should look like, as we shall see. Imagine first what it would take, empirically, to see the same environmental policy implemented all over the world. It would require identical political systems for decision-making, identical regulatory institutions for implementation and identical cultures and perhaps even identical individuals for support and cooperation; all present differences would have to be neglected or eliminated. Moreover, we would all have to be confronted with the exact same environmental problem or same set of equally serious problems – and the same causes – and we would all have to interpret them in the same manner.

Apart from these perhaps too trivial factors, there are a few more fundamental reasons for the existence of 'policy diversity'. For an environmental policy to be effective, it has to meet all sorts of criteria that do not necessarily relate to purely environmental effects. There must be criteria for success, it must be possible for a policy to meet these criteria, it must be possible to meet its goals efficiently, either more efficiently than other alternatives or efficient enough (rise above a certain minimum level of efficiency). Moreover, a policy must be developed on the basis of reliable information – not only about the environmental problem itself and its physical 'environment' but also about the political, legal, social and bureaucratic environment. This in turn implies that standards for the adequacy of information are needed, as well as a decision rule for the choice of appropriate policy instruments and a decision rule for calling a risk (un)acceptable.

Behind all this is another complicating factor: political feasibility. We may question whether existing political institutions are fit to transform a society into a sustainable society, but whatever institutions we put in their place, a policy cannot succeed if policies, institutions and whoever designs or implements them are not recognized as legitimate. Society is like a small ecosystem within the larger global system, and politics in turn is an ecosystem within society. All have their own laws, nearly of nature, the foremost often being that what matters most is the immediate survival, on a day-to-day basis, of individual politicians and citizens.

The means and ends of policies are normative things, and our world is one of moral pluralism. Even if we agree on facts, it is unlikely that we will always agree on moral issues. Hence, there can be disagreement on all the moral factors that determine whether an environmental policy is good: on how to assess risks (as already noted earlier), on what kind of society we want, on what kind of environment we want, on how sustainable we want it to be and in which respects. This is not to say that political or ethical disagreements will always lead to different policies (the means to two diverging ends can occasionally be the same) but neither is there reason to assume that agreement on policies will always go hand-in-hand with disagreement on purposes. Quite the contrary.

In short, there are two main reasons why environmental policies can and probably will differ from place to place even if the problem they address is the same: lack of knowledge and the need for more certainty, and moral pluralism. All this is not to say that there are no drawbacks to policy diversity. It makes life more complicated. On a more serious note, policies can contradict one another or counteract and deviating policies may be immoral from an outsider's point of view. Yet regardless of whether policy diversity does exist or will exist, and regardless of whether a particular instance of policy diversity is justifiable in particular circumstances, policy diversity is a morally good thing, at least in principle. Let me mention three reasons.

First, there are the learning effects: we may expect a diversity of policies to produce more knowledge about the efficacy and efficiency of policies, even if both means and objectives differ. Policy diversity also makes better use of human creativity. Policies developed 'within' one culture or polity will not be too creative or unconventional; yet what happens in country X may well be something we in Y perhaps would not have thought of. An example is the use of free-market solutions in the USA,[39] a solution Europeans would still find hard to come up with. Incidentally, policy diversity also spreads risks. If policy-makers can resist the desire to copy one another's policies, the damage done by an unsuccessful policy can remain limited. There is a similarity here between biodiversity and policy diversity. It is unwise and dangerous for a species' chances of survival to overspecialize or, by analogy, to reduce the numbers and kinds of policies. Diversification is indispensable in the pursuit of knowledge and, hence, in the survival of humankind, particularly of humans in a free and equal society. Policy diversity can also contribute to sustainability on a far larger scale than ecological sustainability alone: in the given context of a region, nation, whatever, it is far more likely than uniformity to produce optimal strategies for cultural, economic, social and political sustainability. Note the similarity between biodiversity and policy diversity: the more diversity (or so the theory goes), the greater the chances of survival in and adaption to changing circumstances. Policy diversity should remind policy-makers and environmental scientists of the existence of moral uncertainty and bind them to impartiality about lifestyles and visions of sustainability.

Lastly, whereas the former argument focused on the good of political or at least practical feasibility, there is also a purely liberal argument in favour of policy diversity: it allows a multitude of conceptions of the good life and of the good environment to be realized, thus offering room – more room at least than uniformity – for the recognition of value pluralism and the realization of individuals' equally worthy plans of life. Note, incidentally, that there is a difference between practical feasibility and morality: the most or only feasible practical environmental policy does not have to be the best one from a moral point of view.

In short, policy diversity is good if and only if it is an adequate reply to uncertainty and to moral pluriformity, one that respects rather than restrains moral diversity. From the individual's point of view, theory of the good, plan of life and conception of sustainability, policy diversity may not always be maximally effective and efficient. Yet if we are prepared to admit to being fallible and to not having exclusive access to some ultimate truth, it is, to quote a famous closet liberal, a good thing that 100 flowers blossom.

8.4 The shape of things to come

If environmental policies can and should differ, then so can and should forms of government and, on a broader scale, societies. Obviously, the easiest way to try to realize sustainability in liberal democracies would involve a central government, licenses for each act, large-scale control, and a veto for every citizen who feels their unconditional rights are being threatened. The ropes do not have to be that tight; liberalism and green concerns are also compatible under circumstances less similar to a nightmare version of social democracy. Environmental control is required only at the fringes of civil society: that is, in the processes of the acquisition of resources and of waste management. Control does not necessarily require policing; in a libertarian society, one individual's protest against an infringement on their rights (not to be deprived of health, a future, a for them insubstitutable species) is the equivalent of a veto in a centralized democracy. Depending on how liberal-minded and grown-up we expect citizens to be, any form of liberal society offers room for sustainability.

The introduction of green concerns into liberalism by means of the restraint principle undeniably brings limitations of individual freedom, but these are the kinds of limitations that are – at least I argued they are – compatible with and even implied by liberalism itself. They limit freedom to the benefit of contemporaries, future generations and animals. Yet since the aim of liberal democratic institutions is to protect our chances of survival under decent circumstances, decent having been defined in terms of liberal concepts of basic needs and further wants, the introduction of these new limits to liberty became rational.

In the Introduction to this book, I referred to the existence of a gap between liberals and greens, more precisely between green ideals and liberal democratic theory, the first being result-oriented, the latter fixated on just procedures. It seems that only a minority among green political theorists believe that procedural, i.e. rights-based, strategies can further the aim of sustainability (e.g. Eckersley 1996; Saward 1996). I think I succeeded in showing the majority view to be wrong, at least in principle. Greens will find it hard to stick to procedures if they do not bring them the results they want them to bring. In this sense, liberal democracy and ecologism are contradictory. But it does not work the other way round. Liberals are interested in procedures (not obsessed by them) but only because these procedures realize something that is worthwhile in ways that are morally acceptable (cf. the stringent criteria of liberal democracy given in Chapter 1, and cf. Barry 1995a). There is, then, an element of endism in liberalism. Better still, sustainability is a valuable goal for liberals and the restraint principle seems to do exactly what greens would want to see happen: it makes a healthy and sustainable environment a cornerstone of society.

That some green political theorists will still not agree to the compatibility of liberalism and ecologism has two reasons. First, greens may sometimes misrepresent liberalism as obsessed by procedures – we would in turn misrepresent greens if we assumed that they would accept the kind of protection of nature that liberals could offer. Assuming they accepted the thesis that there is no difference in practice between a liberal pathocentric interpretation of the problem of sustainability they would still reject it because it does not recognize, as a matter of principle, the intrinsic value of all of nature. Second, they might object that there is more to a sustainable society than theoretically just principles, and the green version of liberal democracy offered here does not in any way guarantee that the sustainable liberal democratic society will be a greener one than what we have now – it may even completely lack the colour green. They would be right on both counts.

Liberal democracy is able to develop policies aimed at sustainability but unable to guarantee that the aim will actually be achieved. It can neither control demand nor supply; the latter is unpredictable, the former uncontrollable. Whether the possibility of sustainability can be realized depends almost totally on the behaviour of individuals in society. If they would reduce their number and change their preferences, the kind of policies liberalism allows will lead us back to a Garden of Eden, forwards to a worldwide Yellowstone Park or sideways to William Morris's dream of England as a perfect garden. If they do not or cannot, the restraint principle allows liberal democratic societies to sacrifice nature, even the lives of animals, bit by bit to satisfy the basic needs of humans. No matter how restricted is our definition of 'basic needs', and no matter how imaginary this scenario may be, the liberal interpretation of sustainability allows that if the survival of humankind so requires, the world is slowly transformed first into a global copy of The Netherlands – cities, grass, cows, greenhouses, grain and nothing else – and then into a global Manhattan without the park, fed on Soylent Green's[40] synthetic food.

What must be noted here is that this green nightmare – it might still be a very agreeable world to live in even for most present humans – cannot be blamed on liberal democracy, nor on the green amendments made so far. It is the logic of the concept of sustainability that allows a near-infinite growth of the needs of the human population. Institutions, liberal democratic or not, cannot ensure sustainability, let alone the existence of a world that is greener than bare sustainability demands. Without the right kind of humans to support them, people without the – in deep green eyes – morally wrong way of life of present humanity, no political system can ever ensure sustainability. If there is any reason why liberalism cannot be green, it must have to do with society, not with liberal institutions.

9

Beyond sustainability

9.1 The limits of sustainability

At the beginning of this book, I introduced three sets of assumptions. One related to a distinction between the political sphere and (the rest of) society. The political sphere was assumed to be concerned with constitutional decision-making, i.e. determining the borders and rules of interaction between and within other spheres in society. It also enforces these decisions. The second concerned an analytical distinction between the political process as a machine transforming preferences into rights and politics as a dynamic deliberative process, in which preferences continually change first in the course of, and due to, the public debate on a political issue, and second in response to the political decisions made as a result of public debate. Although the latter model definitely reflects reality in more detail and more realistically, I opted for the first model as containing all the relevant and no redundant elements. The criterion for relevancy originated in the third set of assumptions: that there are certain criteria, certain moral principles, that liberal democratic constitutional decision-making has to satisfy if it is to count as liberal democratic at all. To establish if liberal democracy can be 'green', it is these principles we have to look at in the first place; if it cannot, there is no need to look any further – liberal democratic polities could then only be green by sheer accident. Hence, I decided that we could make do with a slimmed down, mechanistic model of politics.

The hypothesis was that liberalism, a shorthand term for the moral criteria liberal democracy has to satisfy, can be green. This hypothesis has been confirmed. In this final chapter, I first set out the reasons why I believe in its confirmation – and then I shall try to discredit it again. After all, having established that there can be such a thing as green liberalism is like establishing that we could all have a purple skin. It is theoretically possible but there may be factors other than the human constitution or the compatibility of green and liberal principles that inhibit their becoming reality.

I discuss three possible obstacles for the viability of green liberalism, hoping that they reflect the deepest concerns of potential critics. In Section 9.2, I discuss what may well be a terminological problem: green liberalism as described here may not be green enough to count as green in the eyes of established green political theorists. I shall therefore look at the possibilities of (1) demanding more of liberalism than distinguishing between needs and wants, applying the restraint principle and attempting to ensure sustainability in the weak sense of the Brundtland Report; and of (2) relaxing the stringent conditions of liberal democracy so that short-term human interests have to give way to long-term environmental benefits more than they do now.

A more serious source of concern is economic liberalism (Section 9.3). In the real world and in green political theory, free market conditions, capitalism, trade and commerce, economic growth, private ownership and related institutions and practices are often seen as obstacles for the success of green policies; here I shall consider the possibility that economic liberalism counteracts political (green) liberalism.

Last, I turn to the issue of the genesis of preferences (Section 9.4). Liberalism is occasionally accused of accepting individuals' preferences as given, unquestionable and sacrosanct. Hence, it will supposedly allow the egoistic preferences of individuals to take precedence over the common good of a sustainable environment or – depending on the critic's point of view – it will inhibit the development 'within' individuals of eco-friendly preferences, plans of life, theories of the good and lifestyles. I shall consider two ways of 'improving' individual preferences and thereby the chances of greening society (any society, not necessarily a liberal democratic one): improving democratic procedures and directly influencing people's preferences. My discussion of the first of these two approaches will also show that using the dynamic 'deliberative democracy' model would have added nothing essential to our analysis. I end this chapter and book with a personal note (Section 9.5).

Liberalism can be green because it can address all the relevant topics of debate in green circles, because it can answer them and because the answers are satisfactory. This does not mean that liberal answers will always be orthodox green ('deep' green) answers, nor that they are hardly green ('shallow') yet as green as liberalism can be – it means that those unorthodox liberal answers which at first sight may seem shallow are as 'deep' green as can be desired from any reasonable political theory and that, once put into practice, the world will be as green as any deep green theorist could reasonably desire.

To validate this claim it would be necessary to repeat each and every issue and possible position in the green debate (see Chapter 2). This would be rather fatiguing. Instead, I shall give a short and highly selective sketch of some of the main theses defended in the preceding chapters, sufficiently complete to make my claim plausible. I first deal with ethics and rights, then with politics, policy and sustainability, and finally turn to metaphysics.

If nature or parts of it have intrinsic value, we would have a prima facie reason to respect nature and not interfere with it merely because we could use it in our own interest. In Chapter 4, I argued that intrinsic value, however popular in environmental ethics and green political theory, is a confusing and confused concept. There is no way to get rid of the fact that value originates in a valuing subject, and although this subject does not have to be human, the implication remains that value is external to the valued objects. Yet the damage this does to the green ambition of protecting nature against destruction can be repaired. An object having external value does not mean that it is necessarily substitutable by any other object; it is always possible that it is completely insubstitutable. All it takes is a subject for whose plan of life or theory of the good the existence of this object is essential. Since we owe subjects, i.e. other people than ourselves, respect equal or in proportion to the respect we want for ourselves by virtue of the traits they share with us, this gives us a first alternative prima facie argument against interference with nature.

The traditional philosophical distinction between needs and wants permits us to add a second barrier. Regardless of the way in which this distinction is given shape – it is almost as controversial as the concept of intrinsic value – it allows us to say that there is a difference between unconditional rights (to goods of the needs category) and conditional rights (to goods of the wants category). Arguments for the attribution of rights to needs goods can as a rule only support user rights, not ownership rights: that is, the right to destroy an object can only be part of a person's set of rights if destruction is a necessary condition for its being used. On this basis, I argued for the validity of the restraint principle: no part of nature should be destroyed unless necessary, in which case it should be renewed, replaced or substituted by an adequate compensation. The restraint principle is a second prima facie argument for the protection of nature, one that to me seems far more transparent and intelligible than intrinsic value.

Green political theory has an empirical component next to this ethical one: apart from its critique on human values and attitudes, it argues that there are physical limits to our options for exploiting nature. To ensure that human society becomes sustainable, there is a practical need to limit the demands we pose on nature, to (re)distribute power, opportunity and wealth in order to lessen the pressure on nature, and/or to improve the ways in which we use whatever nature supplies us with.

In preceding chapters, I have taken a close look at each of these types of solutions and drawn several conclusions. First, I – implicitly – argued that sustainability is a rather fuzzy concept: it cannot even theoretically be guaranteed for eternity. It would be better to conceive of it in terms of maintaining a balance between the renewal and the depletion of resources for as long as nature allows. Second, I argued that we should not expect too much from distributive solutions: although redistribution between generations, polities and species for the sake of environmental justice may be required, it can only temporarily reduce pressure on the environment. Only perfect control of either the supply or the demand side of the problem can ensure sustainability over a longer period of time. Third, I argued that demand-side control (read: population policy) is impermissible in liberal democratic societies. Population policies will always lead to the rapid reproducers paradox: that is, to punishing and rewarding exactly the wrong people. Fourth, I discussed the limitations of supply-side solutions and concluded that although they cannot be missed, they cannot be relied on either. In short, liberalism allows many, sometimes radical, policies aimed at sustainability, yet cannot ensure sustainability. On the other hand, neither can any other political theory. The fundamental difference between liberalism and (many) other political theories, particularly green theories, is that it cannot allow population policies but must depend on the voluntary support of individuals. It is only in this respect that liberalism *can* perform worse than other theories; it does not *necessarily* do so – particularly not if non-liberals would accept that the rapid reproducers paradox is an impermissible consequence of population policies in any society, not just a liberal one.

As regards policy, green liberalism offers an extensive set of feasible options rather than one single solution to the question of reconciling liberalism and ecologism. It is in principle compatible with the almost omnipresent government interference in social democracies, with a nearly libertarian night-watchman state and anything in between. Furthermore it takes account of human fallibility and moral discord, thus favouring policy diversity and the implementation of policies on a small scale – for reasons other than, but with effects similar to those of, the precautionary principle. In the end, it supports the creation of a sustainable society without committing itself to a unique road to salvation or to a blueprint of the one and only perfect green society.

Finally, as to the more practical issues on the green agenda – pollution, desertification, extinction of species – I think I have done all a theorist can do: I have discussed principles that can serve as guidelines for real-world strategies, and I have occasionally spelled out the consequences of these principles. In short: in environmental ethics, liberalism replaces intrinsic value by typically liberal notions, yet in practice works out at least as green as traditional green theory; in politics, its position generally corresponds with that of greens; in policies, it is as sensitive to problems of feasibility as moderate ecologists are, and more sensitive than their radical 'Utopian' associates.

One dimension is absent in this overview, as it has been all through this book: the positions liberals would take on issues of green metaphysics. I have ignored this dimension because I was interested in ecologism and liberalism as political theories, and there is no necessary connection between positions on ethical, political and policy issues on the one hand and metaphysics on the other. Yet positions on the former issues do make some positions on the latter less plausible (less compatible) than others. Just for the record, here is what liberals would presumably say on this subject (cf. Chapter 2 for an explanation of the terms used).

Rational liberals are most likely to be 'pragmatic' metaphysical pluralists; to them, it does not matter out of which substance(s) the universe is made as long as it is a conception that allows as many human theories of the good to be at least tolerated and taken into consideration in political decision-making. Since there are people for whom, say, trees and mortgages are insubstitutable and for whom they may be insubstitutable because they can in no possible way be compared on the same scale – it is simply more prudent to accept the possibility of metaphysical pluralism. For virtually the same reason, but now because there are people for whom things can be substitutable, liberals are more likely to think of nature as made up of distinct elements ('compartmentalism') than as a whole ('holism'). Not every butterfly causes a tornado. They will be voluntarists and idealists rather than anything else, not for liberal reasons but simply because they must, after all, believe in the possibility of human autonomy to make an activity like politics worth while. They will and must accept that nature is not naturally in a state of equilibrium – again not for liberal reasons but because there is physical evidence for nature's being continually changing with or without human interference, and because notions like 'equilibrium' and 'keeping the balance' (central ideas for theories of sustainability) can still make sense within an 'evolutionary' framework. Finally, liberals will or should be organicists rather than mechanicists, the former position allowing them to be more sensitive to the dangers of human fallibility in harmonizing human demands with nature's laws.

So much for the good news: liberals can, indeed they must, be green, and although nothing obliges liberals to revere Mother Nature, repent and sin no more, liberalism itself obliges them to respect the views of those who do feel this way and to act in ways that preserve nature as much as possible, as much as deep greens could reasonable demand. Yet principles are easy; implementing them is another story. If liberalism is so green, then why are really existing liberal democracies facing environmental problems, even, if we may believe the doomsday sayers, an environmental crisis? And how is it possible that green liberalism, the theory *itself*, does not straightforwardly exclude the possibility of turning the planet into a giant steel-grey Manhattan? In the next three sections, I shall answer these complaints.

9.2 Relaxing conditions, strengthening ties

A first possible reason why green liberalism does not appear to be as green as ecologists would want it, or any political theory, to be is that political institutions – as distinct from the rest of society – should do more to count as green than promoting a more economic way of exploiting nature only. Thus, we either formulated our goal (sustainability) too leniently, or the strict conditions of liberal democracy are too strict, or both.

Let us start with sustainability. The Brundlandt conception of sustainability used so far is, admittedly, merely a description of what a sustainable human society should look like; all it does with regard to a sustainable *environment* is formulate a few minimal conditions, very permissive side-constraints. It is this permissive character that forced us to conclude that a liberal democratic sustainable society could take any shape, up to and including that of a global Manhattan. Green theorists from Næss and Baird Callicott to Barry and Dobson, on the other hand, would argue that the *raison d'être* for ecologism and environmentalism is a suspicion that we are presently reaching the limits of how much exploitation and destruction nature can stand, either locally ('environmental problems') or globally ('ecological crisis'). If, on these views, a conception of sustainability allows the permanent deterioration of nature rather than at the very least demanding that things do not get worse – then there must be something wrong with this particular conception of sustainability.

To improve Brundtlandian conceptions of sustainability accordingly, special demands should be added to ensure environmental sustainability independent of but analogous to human sustainability – something along the lines of 'fulfilling the present needs of non-human nature without compromising chances for their fulfilment in the future'. Unfortunately, an amendment of this kind will get us into trouble with either human sustainability or liberalism. If the demands for human and environmental sustainability contradict one another – a not too unlikely scenario – at least one of the two has to give way. If we give priority to human sustainability at the expense of environmental sustainability, we will be back where we began: a room with a view on the global Manhattan. If we give equal priority to the two, both will have to give way. Both in this case and in case we prioritize environmental sustainability, human sustainability will be compromised. The least that will be required for environmental sustainability here is zero growth, either in the use of natural resources or in the size of human demand. Limiting demand will, as we saw in Chapter 6, bring us into conflict with the stringent condition of liberalism on the one hand and on the other implies that the satisfaction of the desires of present humans becomes impossible – which violates a criterion of human sustainability. We can exclude this option for now. Limiting the use of resources beyond what is already required by the restraint principle means that it will not just be the desires of humans that will remain unfulfilled (a violation of human sustainability) but that even their needs will be neglected – which goes against the grain of liberalism. In conclusion: if environmental sustainability is to be guaranteed at least to the same degree as human sustainability, either the present criteria for human sustainability will be violated or those of liberalism.

Note that this conclusion could be reached without (re)introducing notions like intrinsic value and future generations. First, the possibility of a conflict between environmental and human sustainability is one that concerns existing human beings; sacrificing their interests to environmental sustainability may (but does not necessarily!) lead to the creation of a population in the future whose (size and) demands do not conflict with environmental sustainability – the point being that we *do* have to sacrifice the prospects and well-being of presently existing humans. Second, reintroducing the idea that nature has intrinsic or independent value would only have complicated matters further. Either all of nature can used by humans despite its having intrinsic value, in which case the practical difference with our set of alternative concepts (external value, (in)substitutability and restraint principle) collapses, as concluded in the previous section, or the concept of intrinsic value is interpreted in such a way that (1) all parts of nature have equal value or (2) some have more value than others and are put beyond the reach of human exploitation. In the first case, intrinsic value offers no guidance whatsoever – killing the last gorillas would be as immoral as eating carrots. In the second case, we will reach the same conclusion: human sustainability has to give way to environmental sustainability.

If we can improve our conception of sustainability at all so that human and environmental sustainability need no longer conflict, we can still get into conflict with the stringent demands of liberal democracy. An amended interpretation of sustainability that allows demand-size control directly violates liberalism; one that allows supply-side policies will violate liberalism if it goes beyond the demands of the restraint principle. In short, to exclude the global Manhattan scenario those liberal conditions that preclude population policies or further economizing have to go.

Since I discussed these two strategies of sustainability in Chapters 6 and 8, I can be brief here. As to population control, I argued that three of my stringent conditions bar liberal democratic regimes from developing population policies: proportional equality, the liberty of life and the deliberative requirements posed on collective choices. The last of these three must be overruled by the other two, unless we want other liberal principles to become matters of convenience: after all, under the right circumstances it can be quite reasonable for democrats to curtail liberty and decrease equality, just as it can be quite reasonable for a utilitarian to admit that slavery can sometimes increase total welfare. In other words, giving priority to democracy, deliberative or not, would take the 'liberal' out of liberal democracy. Adapting the proportional equality condition so that (the parents of) 'sustainable' children deserve protection against (those of) 'superabundant' children cannot be done without the possibility of arbitrarily violating individuals' liberty of life: the superabundant j may, after all, be far more important for its parents than an accidentally created i for its parents. Finally, limiting the liberty of life itself turned out to be a hopeless strategy. The antinatalist would either have to argue that procreation should be limited despite the fact that reasonable procreating individuals have good reasons to procreate, or that it should be limited because the individuals in question are irrational. In the latter case, the link with environmental concerns would be lost; in the former, environmental concerns would be allowed to overrule even the most basic needs of individuals, once more taking the 'liberal' out of liberal democracy.

Consequently, liberal democracies cannot as a rule be permitted to control population growth – there may be room for policies aimed at preventing procreation in individual cases, ranging from rape to teenage pregnancy, but these cannot be considered macropolicies, let alone policies inspired by environmental concerns. Preventing overpopulation is the responsibility of individual citizens. Alternatives to enforced macrolevel birth control were also excluded, again with exceptions: the restraint principle does allow liberal democracies to put limits to the satisfaction of non-basic desires (wants). Decreasing welfare to create poverty, leading to smaller humans with smaller needs, was considered perverse; authoritatively changing people's diets or appetites was condemned as interference in the private sphere. In the end, then, people's needs – for company, children, food, travel, technology and trinkets – are private affairs; control, if possible at all, is impermissible. All that a liberal democratic regime can do to check population growth is enable citizens to make a choice between procreating or not and *inform* them about the possible consequences – but it cannot be permitted to cross the line between information and indoctrination.

There was, as I showed, more room for supply-side solutions – in fact, from a deeper green point of view, too much. I did not only allow techniques and technologies that made better use of existing resources, technologies that might even reduce pressure on the environment, I also allowed an actual increased use of the environment in the name of human sustainability. The proponent of environmental sustainability would argue that preventing this latter option is a necessary condition for environmental sustainability (read: conservation or improvement of the present state of nature). Unfortunately, the only way to exclude the further exploitation of nature is by mutilating the liberty of life. Once the needs of a human society increase to a level where they cannot be satisfied without new resources (or dangerous technology), we would have to permit that at least some humans live a life they do not consider worth living. In fact, we would not just permit this but actually compel people to live a miserable life by failing to do what we can do: that is, make the resources that they need available. This, it seems to me, will do more than take the 'liberal' out of liberal democracy. It would take the humanity out of it.

There are definite limits to how far the stringent conditions of liberal democracy can be relaxed. It appears to be impossible to bend them far enough to avoid a deterioration of the environment under any and all circumstances. Theoretically, a global Manhattan remains possible. The essence of a theoretical possibility is, however, that it can only occur under specific conditions. If we want to prevent it, we shall have to look outside the realm of constitutional principles and ask whether the conditions under which liberal democracies would be forced to live with this consequence can be precluded in another way. As I suggested in the previous section, there seem to be two other possible causes – and therefore two possible cures – for the deterioration of the environment: economic liberalism and individual preferences. Let us consider these next.

9.3 Economic liberalism

The core argument of this book has been that both political liberalism or liberal democracy, and philosophical liberalism, its theoretical foundation, have a potential for becoming green. Neither one is therefore a necessary condition for the existence of the environmental crisis – if there is such a thing. Green political theorists, however, have asserted that there are other causes of environmental problems next to liberal democracy – or instead of liberal democracy, since some would agree with my argument, one way or the other (cf. e.g. Achterberg 1993; Jacobs 1994). In previous chapters, I have already discussed two of these possible causes, technology and overpopulation; this section deals with economic causes. In the next section, I shall discuss a fourth and last set under the heading of 'consumers' preferences. It is important to stress that liberal democracy and economic liberalism are neither identical nor Siamese twins. Capitalism is not a necessary condition for the existence of the political structure and liberties associated with liberal democracy (cf. Lauber 1978), the free market existed before liberal democracy evolved, political liberalism is not predestined to a *laissez-faire* attitude towards the market (cf. Stephens 1996), nor does it hold that life is all about making profits or the satisfaction of material self-interest; it sees trade and commerce as one way of life, one means to the realization of a plan of life, among others (cf. Holmes 1993: 212). The critique of economic liberalism must therefore be judged on its own merits and cannot reflect on political liberalism.

What should also be remembered is this: there is no such thing as a society without a free market. Not only in planned economies, but even in the direst of times and under the most oppressive systems of government, something called 'the black market' operates – and it operates on the same principles as the free market: demand and supply. In fact, the only difference between the black and the free market is that the latter is legally permitted. The argument I make, then, is that if 'the' free market is to be blamed for environmental problems, it cannot be blamed because of its being a *free* market; it is because there are consumers and producers who are looking for a market – any kind of market.

Greens blame either economic liberalism or industrialism, the second of which refers to the economic systems of both the liberal democratic West and the former communist East (e.g. Bahro 1980). Analyses in terms of industrialism interpreted state socialism roughly in two ways. On some accounts, it was a pendant of capitalism, a system that was forced to compete on all fronts with the one it was meant to replace; on others, it was predetermined to accommodate the same misguided interests of consumers and (most importantly) producers that reigned capitalism. On all crucial points, the first type of analysis can be seen as a special case of the critique of economic liberalism and the second as critique of individual preferences. I shall therefore, at the risk of being incomplete, ignore both in these two sections and focus on the 'Western' case only.

One of the odd aspects of the green critique of economic liberalism is that some see it as the source of most environmental evil, whereas others believe that it is the solution. 'Leftish' greens, who on the left side of a map of the world are usually seen as the more truly greens, argue that private ownership and free market conditions in general, or capitalism and the perpetual quest for economic growth in particular, are at least barriers for a greener society, often a cause of environmental degradation and according to some even the main cause (e.g. Smith 1995). On the right, especially in libertarian circles, it is state interference that prevents the free market from developing into a green market (Anderson and Leal 1991, 1994; Narveson 1995). If property and use rights were better defined, that is, attributed to individuals rather than left unattributed by and reserved for the state, individuals would be enabled to defend their rights against polluters and exploiters of (their) nature – and the mechanism of the free market would ensure that no more nor less natural capital than necessary were used for consumption.

Although most of my critique is directed against the first, 'leftish' perspective on economic liberalism, I hope it will be or become clear in the course of this section that I also see serious problems in the second perspective.

The first and most important reason why economic liberalism is suspected of causing environmental damage is its expansive nature, its pursuit of economic growth (e.g. Achterhuis 1994; Opschoor 1994). Now it should be noted that there are two kinds of economic growth that are not, or not significantly, environmentally unfriendly. First, imagine an eco-farmer entering the free market and discovering not just a niche there, but actually immense opportunities for growth. Although her prices are higher and the demand for vegetables does not increase, she attracts most of her pesticidal competitors' customers and, within a few years, she has conquered the greater part of the market. The volume of the market is the same, the prices are higher, therefore the GNP has grown – while environmental conditions have improved. Replacing environmentally damaging products or technologies by more benign ones (a cornerstone of the ecological modernization theory, cf. Weale 1992) allows economic growth without environmental damage. A similar observation applies, second, to a society of three people where A and B grow food and sell it to C for £100 once a year, while C makes and sells all sorts of utensils, also for £100 annually. Now A and C hire B for administrative work, each paying B £50 – in the old days, B did this for free. The GNP has grown by 50 per cent without increased damage to the environment. Not all economic growth is material; it follows that there *is* such a thing as sustainable growth (*pace* Achterhuis 1994).

The size of the economy as measured in money, then, says little about the environmental effects of economic growth. It is either the *type* of growth that matters, i.e. whether or not it increases the pressure on the environment, or the *tendency* for growth as such. Let us pursue the first possibility first.

Assume that a company expands – it builds a new factory – and that its growth increases pressure on the environment without actually destroying our collective chances of survival. The company's products increase our welfare, its expansion decreases it. Two comments are now possible. First, the critics remark that economic actors do not usually take account of the environmental costs they (make or) cause; economic liberalism allows them to ignore these costs and burden taxpayers and future generations rather than themselves and other beneficiaries, the consumers (Opschoor 1994; Narveson 1995; Smith 1995). Apart from the fact that this is an empirical issue rather than one of principles, since there are practical ways of taking account of environmental costs (taxes, audits) – the critique misses the point. There is no reason why free enterprise should be the only system that allows actors to sweep environmental costs under the carpet: the same can happen, and actually happened, in countries where 'collective' ownership predominated. A libertarian (Anderson and Leal 1991; Narveson 1995) would even argue that if individuals rather than the collective owned the environment, there would be less room for producers to get away with environmentally harmful behaviour: the state is often blind, individuals on the other hand will take the greatest care that their bit of the environment is not damaged unrightfully. Although I agree for technical reasons that liberal democracies should attribute all possible rights unequivocally and that they should be attributed to individuals rather than collectives (cf. Chapter 5), I would not want to support the libertarian claim that there is too little free market. There is no reason to assume that an individual actor will not be prepared to, say, sell a polluter the right to pollute their grounds and fatally injure a particular biosystem if that is profitable for them, individually and in the short term. Nevertheless, the fact remains that there is no necessary link between the free market and evading environmental costs.

A second possible comment is that damaging the environment for the sake of material welfare is, or under circumstances can be, immoral. In previous chapters, I have argued that there are circumstances under which this is not the case; I do not want to repeat myself here. The problem with this kind of argument, even if it were valid, is that once more it has nothing to do with economic liberalism. It applies equally well to state capitalism.

Hence, if there is something wrong with the free market and private ownership, it must be that the system has an unjustifiable tendency for growth built into it. Whereas (ideally) collective ownership of producer goods and collective decision-making on production targets would result in supply meeting needs as efficiently as possible, private ownership forces economic actors to increase their profits and their share of the market, even if that implies having to create new needs or supplant (environmentally) superior products and techniques by inferior ones. (Incidentally, a critic of the Soviet system would remark that it is precisely the absence of incentives for innovation that contributed to the collapse of the Russian economy – cf. Weenink 1997.)

It is probably true that capitalism as the youngest incarnation of economic liberalism, or even economic liberalism as such, only helps those who help themselves: that is, that it rewards both necessary and unnecessary innovation regardless of its environmental effects. The entrepreneur who does not follow his competitors in introducing ever more 'improved', ever better looking, ever flashier advertised washing powder will loose customers. The free market can operate as a perverse cooperative game:

Table 9.1 Rewarding innovation

	B: cooperate	B: defect
A: cooperate	(3,3)	(4,1)
A: defect	(1,4)	(2,2)

In a two-actor economy, if A initiates an innovative strategy (introduces a new type of washing powder), and B does not, their gains are (4,1): A wins almost the whole market, B gets the worst possible results. If B innovates and A not, A loses. If both refuse to innovate, they will still be worse off than when they both innovate. The rational thing for them both to do is to innovate.

Yet it is not the free market that should be blamed for effects of this sort. The free market does not create the structure of rewards: consumers do. The free market does not make one result more desirable than another: the producers do. In one alternative world where consumers would not fall into the trap of un-necessary or environmentally harmful innovation, the reward structure would look differently – defection would actually be rewarded. Such an aggressively conservative market is not a fiction – a fact to which the Coca Cola Company can testify ever since it made business history with its catastrophically unsuccess-ful attempt to change the classic flavour of its product (or the taste of its clients, or both):

Table 9.2 Punishing innovation

	B: cooperate	B: defect
A: cooperate	(3,3)	(1,4)
A: defect	(4,1)	(2,2)

In a second alternative world, the producers would find it rational to aim for the highest possible gains but be satisfied – for reasons that have nothing to do with profit growth – with the results of defection, thus changing the matrix into:

Table 9.3 Discouraging innovation

	B: cooperate	B: defect
A: cooperate	(2,2)	(4,1)
A: defect	(1,4)	(3,3)

I shall skip the mathematics (I would not want to make this a boring book) and immediately move on to the conclusions that could be proven with the help of a deeper analysis of free market games. There are many more alternative worlds imaginable in which only a part of the consumer population would reward innova-tion or where only a part of the producers are more interested in the rewards of innovation than in those of defection. Some of these worlds are in equilibrium: that is, the (non-)innovative producers can serve the market of (non-)innovative consumers. In nearly all these worlds, the consumer who is interested in 'new' brands of washing powder may be immune to the seduction of 'new' audio equip-ment and vice versa. To complicate matters even further, since trade is an iterated game, we can imagine worlds in which the environmental effects of the non-innovative markets compensate for those on the innovative markets. In a society of this type, the free market can actually be environmentally sustainable without anyone being forced to change their preferences. I would not go so far as to suggest that this is a necessary consequence of a free market – and I really would like to see the libertarian argument for it – but the point to note is that environmentally harmful effects of economic liberalism are ultimately the result of the preferences of consumers (see next section) and producers (see below).

As a matter of fact, the free market is not a necessary condition for the existence of an environmentally perverse economic rewards structure. Provided we are talking about a society inspired by the ideal of being responsive to the needs and wants of its members, the presence of no more than two economic actors (one buying, one selling) is enough to get the same results: there may always be consumers who crave for the environmentally harmful. Innovative producers in an aggressively conservative free market society will not stand to gain. In fact, we do not even need a consumer-responsive society; it is enough that there are consumers and producers willing to negotiate with one another, be it in a free market society or on the black market. That individual preferences are ultimately the source of all free market evils is true not only for growth-based arguments against economic liberalism but also for criticisms on 'the commercial attitude'. Advocates of this argument refer to economics as a 'disembedded' and 'imperialistic' way of life (Achterhuis 1994) and criticize the tendency of economic actors to commercialize every good, from sex and religion to sunlight and clean air, and then create a market for it. This argument is not an ecological original – it is a classical and medieval theme, also taken up in this century by communitarians like Walzer (1983). Three things should be said about it. First, it is a highly rhetorical argument. If taken seriously, the remark that businessmen can take any object and squeeze money out of it can only be answered by 'So what?' To be a normative statement rather than a statement of fact, one must assume *a priori* that there is something dirty about commerce, either intrinsically or in specific cases. In the first case there must be walks of life that are more honourable or palatable, but as long as a society without trade is unimaginable it merely represents an inconsequential elitarian attitude – and a distinctly anti-liberal attitude if it did have political consequences. In the other case, we need further arguments – real arguments – to show exactly which goods cannot or should not become objects of trade. In our case, we need environmental arguments.

Which brings me to my second remark. As argued before (see Chapter 7), there can be no principled objection to psychologically transforming nature into natural capital – in terms of the anti-trader, to seeing money in everything. It is an indispensable attitude in a world of limited resources and permanent need. Trade being merely a means to satisfy these needs, there can simply be no valid objection to the commercial attitude – even assuming that to be a trait of economic liberalism only. What we might legitimately object to is the transformation of nature into unnecessary goods or the unnecessarily destructive transformation of nature into necessary goods – necessity here is defined as being incompatible with the restraint principle.

217

Finally, the creation of unnecessary goods in the latter sense is in no way typical of economic liberalism. Unnecessary production can occur in any type of economy; it just takes the right (wrong) kinds of individual preferences on the sides of consumers (see next section) and producers.

And so we reach the last argument against economic liberalism: the egoistic accumulative attitude that would characterize the world of free enterprise. There are three sides to this – admittedly peculiar – coin. On the one side, there is the pressure of capitalism forcing economic actors to act in their short-term individual interest and shift environmental costs on third parties. Although this phenomenon is not characteristic for capitalism only but apparently something that can occur in any economic system (see above), it is not insuperable. Either a more libertarian free market or more government interference (or both) can limit the extent to which environmental costs are evaded, companies can be made more aware of their long-term interests (cf. Opschoor 1994).

The second side of the coin is that capitalism means power and big business: it is not about individuals producing and trading goods but about large firms and international consortia. The human actors within these economic actors may identify as much if not more with the welfare of their company than with that of their society. Forcing companies to conform to environmental standards – either those of governments or those of individual right holders – may just be not enough to prevent firms from moving to other countries or to ensure that all branches of a firm and all individuals within them conform to official corporate policy.

The third side of the coin, then, is that institutional arrangements alone cannot ensure that economic actors will conform to environmental standards. The market (consumers) must reward 'good' behaviour and the *individual* productive actors must be inspired accordingly. Being a philosopher, it is hardly my line of work to establish whether there is room in the economic sphere for an environmentally benign business ethics. The willingness of corporate managers to engage in courses on environmental management and business ethics and to take up their public responsibilities should perhaps be set against the business world's proclivity to participate in the most absurd management courses and its consumeristic trust in every new management fad and fashion. All I can do is make it plausible that there can be an environmentally friendly business ethics. As a matter of fact, its basis has already been laid on the one hand by the analysis of fair pricing by Aristotle (1959, 1980) and his followers, and on the other by late medieval political theorists who attributed merchants with a public responsibility for enabling society to find the golden mean between want and extravagance. Allow me to quote one, perhaps not the most brilliant but certainly one of the most readable sources:

> (T)he merchant class is very necessary, and without it neither the estate of kings and princes nor even the polities of cities and countries could exist. For by the industry of their labour, all kinds of people are provided for without having to make everything themselves, because, if they have money, merchants bring from afar all things necessary and proper for human beings to live . . . And for the good that they do for everyone, this class of people – loyal merchants who in

218

buying and selling, in exchanging things one for another by taking money or by other honest means – are to be loved and commended as necessary, and in many countries are held in high esteem . . . These people ought to be well advised in their deeds, honest in their labour, truthful in their words, clever in what they do . . . and ought to be well informed about whether there are enough goods and where they are going short and when to buy and when to sell – otherwise their business will be gone. (De Pizan 1994: 103–4)

Economic liberalism contains many elements that impede sustainability, but none of these threats seems invincible – assuming the right kind of preferences.

9.4 Deliberative democracy and the last taboo

Whether or not liberal democracy can realize its potential for sustainability, whether or not it will be able to prevent society's developing into a global Manhattan and whether or not the sustainable liberal democratic society will offer more room for the free flowering of nature – indeed, whether any political system can do any of these things – ultimately depends on the preferences of humans. In the words of someone who truly believes in an impending ecological catastrophe: 'If we were not to survive the environmental crisis, this is not liberal democracy's fault but our own' (Jacobs 1994: 163).

In the greater part of this book I have taken preferences as given. Political liberalism (liberal democracy) offers some protection in the form of the restraint principle, but no guarantee that the volume of even the most basic human needs will not in the long term lead to the replacement of nature by (to human needs) a 'more adequate' artificial system of resource production. Economic liberalism allows all kinds of preferences without exception; the only perspective it offers for sustainability is the perhaps slightly optimistic expectation that an unrestricted free market and a complete attribution to individual actors of all possible rights to nature will, assuming the existence of rational individuals, lead to nature's being protected as much as the individual owners will and can. There is no guarantee that the interests of animals will be respected, no guarantee that self-interest and nature protection will not conflict when cooperation over the borders of individual estates is required, no guarantee that future generations will not inherit a seriously deteriorated planet.

Grim as this worst-case scenario may be, we should not make the mistake of equating liberalism with liberal democracy or the free market. In philosophical liberalism, preferences are *not* given. Liberal philosophers take a keen interest in the genesis of preferences. First, a liberal democratic system obviously sets limits to the kinds of preferences that can be satisfied; if it is to be stable, individuals must learn to develop the kind of preferences that the system can handle. At the very least, liberals demand limits to the system's tolerance for the intolerant (Holmes 1993: 244); in more elaborate theories like John Rawls's, the well-ordered society

219

requires individuals who give priority to the right over their personal notions of the good – and in due course accommodate their theories of the good to the limits set by the need for social cooperation (Rawls 1972, 1993).

Second, even against this background liberals do not accept every preference 'left' as equally relevant and worthy. It not only questions the admissibility of 'perverse' preferences, the satisfaction of which might be harmful for others, but also preferences originating in so-called expensive tastes. Liberals classify preferences as voluntary or involuntary, reversible or irreversible, based on full information or on lack of information; each of these distinctions may justify the inclusion or exclusion of individual preferences. The distinction between liberalism and other political theories is not that it does not allow a debate on the kinds of preferences that are and are not admissible; it does. The distinction lies in the criteria it sets for calling a preference too 'perverse' to accept – and in the fact that it rejects interference by higher authorities in the process of preference formation.

As is it possible within a liberal framework to discuss the genesis of preferences and to call some 'good' and others 'bad' (or words to that effect), there is no reason why we should not discuss the question whether there are environmental reasons for dismissing some preferences as too perverse, whether the development of more environmentally benign preferences should not perhaps be promoted, and how this could (read: may) be done. I shall deal with these questions in this order.

There are basically two reasons why environmentally harmful preferences can be seen as (too) perverse: either because they are legitimate as preferences for an individual in their individual circumstances but disregard the general interest or the common good, or because they are simply the wrong preferences. An example of the first case could be using pesticides to save our crop; it helps the individual but harms nature and the human public. An example of the second case is an individual who, on the basis of insufficient information or an ethically depraved ideal, sets themselves the task of eradicating, say, all worms. Three remarks are in order here. First, note that both types of reason are compatible with liberalism. Liberalism does not deny that there can be such a thing as the common good (cf. Holmes 1993: 199), first and foremost in the shape of a well-ordered society – it merely denies to have access to privileged knowledge on the one and only substantive *eidos* of the good. It also admits the possibility that some preferences and even complete theories of the good are ill-informed or unethical; there is a distinction between having access to the ethical truth and knowing what can *not* be good. Next, both types of reason can serve as grounds for deliberative democracy, whereas only the latter can be a premise in an argument for a less tolerant, more communitarian alternative to liberal democracy. Last, greens tend to object to real-world liberal democracy's treatment of nature on the basis of the latter argument rather than the former. The ideas that incite environmentally harmful behaviour – materialism, greed, etc. – are seen more as unethical (a 'wrong' conception of the good life or of the value of liberty) than as in principle ethically neutral.

Are these good reasons to not just call certain preferences environmentally harmful but to also reject them? My answer in both cases is a qualified 'No'. As to the idea that people can have a 'wrong' view of the good – of what is good in general or what is good for them – I readily admit that lack of information is a reason to qualify a person's theory of the good and the preferences they derive from it as 'insufficiently justified'. The same is true for a judgement based on false information or for one based on prejudice, irrational likes and dislikes, logical mistakes and the like (cf. Brandt 1979). All these are considered to be obstacles to individual liberty: that is, avoidable limits to the individual's ability to develop their own plan of life. However, this in itself is not enough to disqualify any preference. Consider a down-to-earth example: if my intuition tells me that 523 plus 677 makes 1200, we will not disqualify the answer because I used the wrong method to reach the right conclusion. Incorrect procedures and incomplete information increase the risk of making wrong practical decisions and *may* increase the risk of making wrong ethical decisions – they only *will* if we have a criterion for either ethical rightness or ethical wrongness.

Although some liberals will deny that there are criteria for the ethically wrong, one of the defining marks of liberals is that they will never claim to have criteria for the ethically good. I have already discussed this to some extent in Chapter 3: liberals are sceptics on this point. They will therefore reject any attempt to reject an individual's perception of the good for being 'antithetic to the good'. On the other hand, there is room in liberalism for the argument that some preferences can *not* be good (cf. Raz 1986). If, for instance, freedom for a specific intolerant faith view shared by a part of society would result in (civil) war, genocide and the destruction of the fabric of society then no liberal, believing as they do in the all-transcending value of individual human lives, can avoid the conclusion that (the consequences of) unlimited freedom for this particular faith would be ethically wrong *as well as* incompatible with the idea of the right.

In accepting that some preferences can be ethically wrong, it might seem that I open the door for the argument that it should matter that people are positively free in the classical sense of positive freedom: being one's own master – that they should at least be freed of slavery to wrong ideas. I do not, though. Assuming that the classic idea of positive freedom makes logical sense (*pace* Berlin 1969 and Steiner 1983) it still takes a substantive idea of the good to distinguish the positively free human from the slave. Hence, we will find liberals supporting the idea that a good life can be characterized by the immaterial pleasures of friends, company, games, enjoying nature by day and the stars by night and all the other elements of a (more) pastoral life as propagated by some greens (e.g. De Geus 1993). But we will not find any liberal supporting the view that a more rustic lifestyle is a necessary condition for the good life. Most will concede that if there is such a thing as 'the' good life, it could well be one of luxury and lechery.

With some terminological changes, these same observations apply to the second reason why preferences might be rejected: their being right from a personal point of view but wrong from a social one. First, lack of information is a good but insufficient reason for rejecting preferences. Second, since liberalism does not allow the authoritative prescription of a social ideal, it can only permit exclusion of preferences if they are ethically wrong or imperil the common good of a stable framework of social cooperation – the latter including, as I hope to have shown in this book, sustainability in the form of the restraint principle.

In more precise terms, then, there are three reasons why liberals would accept the judgement that some individual preferences are 'wrong' and why they would consequently allow their genesis to be open to critical examination. The first reason is a technical failure in the process of preference formation: false or insufficient information, incorrect analysis of information. The second is the common good of a viable society. The third is one not all liberals will accept: a preference being evidently ethically evil. The first justifies the promotion of the unimpeded development of my plan of life on the basis of my own interests; the other two point to the interest others have in my free development. They justify the development of individuals in freedom and draw the line where authority comes in.

It should come as no surprise, then, that liberal democracies take an interest in the existence of a public sphere free of government or other compromising interference; after all, this was one of the premises on which the model of liberal democracy as a machine was based. The question is: what should this sphere look like? A complete analysis would discuss liberal (side-constraints to) education, liberal (side-constraints to) friendship, liberal to lover's and family relations, and so on – but for practical reasons I can only examine the most directly political aspect of the public sphere here, the interface between the public and political spheres. One way of ensuring – as much as possible – the unimpeded development of preferences is to promote 'deliberative democracy': a dialogue among citizens and between citizens and their representatives in government in every phase of the process of political decision-making. The liberal theory of deliberative democracy prescribes that some principles are beyond discussion (cf. Cohen 1997: 145, 151; see also Jacobs 1994: 161) to protect the liberal and democratic character of a society against self-destructive decisions; public debate is supposed to reinforce commitment to the basic values of society rather than put them in peril. In the real world, liberal democracies go much further in restraining the impact of public debate, down to the level where the influence of citizens is restricted to elections on a four-, five- or six-year basis, concentrating all political issues in the choice for one party or candidate. Particularly if representatives want to be re-elected – and bow to public demands – this cannot be a sound foundation for a dialogue, let alone for an emancipatory dialogue.

Deliberative democracy is a hot topic in political philosophy these days, and green political theory is no exception (cf. e.g. Doherty and De Geus 1996; Lafferty and Meadowcroft 1996; Mathews 1996). It has even inspired one of those rare experiments in which political philosophers, empirical scientists and politicians cooperated: James Fishkin's project, in which he compared the preferences, degree of information and quality of deliberation of an quarantined group of English citizens before and after he let a group of politicians loose on them (Fishkin 1995). The results he reported are remarkable and a slap in the face of anyone who feels that more democracy necessarily means throwing pearls before swine. Yet putting more democracy into the public sphere has one important disadvantage: it may be emancipatory, it may result in the development of considered preferences, it may increase the legitimacy of systems and policies, it may improve the quality of decision-making and lead to more informed choices but there is no reason whatsoever to believe that it will improve the moral *quality* of the decisions themselves (cf. Barry forthcoming). There is reason to believe that decisions on environmental issues will not be an exception to this rule; in fact, there is reason to suspect that environmental questions are more than averagely vulnerable to substantively 'bad' decision-making.

Democratic decision-making and the creation of a substantively good society are related in the way having a driver's license is related to being bald; the relationship is purely accidental. The differences are that there is a degree of agreement on what constitutes baldness, but not on what makes a society good, and that baldness exists, whereas even the possibility that 'the good' exists is controversial. Perhaps a comparison between the good and searching for a black object that may or may not be somewhere in a totally dark, totally spherical, infinitely large room in zero gravity is more appropriate. In short, there is no reason to believe that even after a process of dialogue and deliberation, any random set of flesh-and-blood individuals will make the good decision. If the issue at stake is of an environmental nature, the odds that their decision will be a disappointment to greens even increase. First, the interests of the environment are not as close to the heart as our more direct interests: in any case where 'acting green' would ask for a personal sacrifice or for a change of preferences, the advantage is on the side of a conservative attitude and it is the change that has to be justified. Second, the same applies on a collective level. More democracy cannot do away with the self-interest of a community, nation or firm. The shared interest (call it common weal, if you like) of, say, the British tribes in a clean and sustainable Britain can imaginably be most effectively satisfied by spreading its pollution in the wind and its waste over the seas – to the disadvantage of foreigners. In both cases, there does not have to be ill will – the problem is simply that in real-world debates, some interests and points of view are less easy to represent and there is no guarantee that all those that matter are actually represented.[41] Perhaps these – not to mention other – shortcomings of democracy can help non-liberals understand why liberals from left to right and from Rawls (1972, 1993) to Nozick (1974) prefer to use fictitious 'reasonable persons' and fictions like a state of nature or the original position as a basis for their political theories.

The green proponent of deliberative democracy may now counter that even if more democracy does not guarantee better decisions from the point of view of environmental ethics, they will be more informed. True. So what? Fictitious reasonable persons or real-life informed citizens confronted with a choice between saving a forest (of which there are still enough) and an endangered but inedible species of snail on the one hand and a road, jobs, factories, income, security on the other will not necessarily choose the snail just because they are better informed.

Facts, no matter how many facts, are meaningless without means of interpreting them – without a moral point of view. Deliberative democracy is a great good, but not because it serves green interests. Public debate may stimulate the development of less environmentally harmful preferences, information may lead to a more responsible treatment of our natural environment, yet it leaves a global Manhattan as a possible world.

There are other ways to promote attitudes favourable to environmental protection, conservation, sustainability and anything greener, more than deliberative democracy can do, but these require indoctrination rather than information. The creation of community feelings and genuine communities with a shared culture is one possibility – a unitary culture makes it far easier to censure and prevent anti-social behaviour; communities offer a stable and secure environment in which individual plans of life do not need to be developed from scratch. Yet if we want the kind of communitarianism that supports and secures a green way of life, there are two options: to pursue a kind of harmony between community and liberal democracy, or to reject the latter altogether. The first alternative is no real alternative: liberal democracy has no principal objections to anything consenting adults agree to, except an agreement to act as if the criteria of liberal democracy did not exist. The first outsider entering a community, the first recalcitrant kid, the first grown-up coming out as a materialist, is enough to change a harmonious culturally monolithic community back into a pluralistic liberal democracy. The second alternative is obviously unacceptable to liberal democrats – but for the sake of completeness I shall discuss it briefly anyway.

It is perhaps a bit cheap to describe fully-fledged communitarianism as another road to serfdom (cf. Hayek 1976), but it is a perfectly appropriate description. Green communitarianism would be the nightmare of Utopia come true. It would be a world of fear – fear for new techniques, developments and ideas, fear for environmental risks and dangers, fear for one's neighbours. To ensure that a communitarian society would conform to a particular ideal of the environmentally friendly sustainable society, important liberties would have to be curtailed. Perhaps freedom of thought and speech would still be formally respected, but social control or in the worst case the *inoffizielle Mitarbeiter* (spies – our children, parents, loved ones) of a benevolent junta would make it difficult for the dissenter or the critic to be heard. Respect for nature would have to be both taught and put beyond question; the liberated but no longer free market would provide a new brand of opium for the people, perhaps Arne Næss's ecosophia (1989). The liberty of lifestyles and with it the liberty of life would be sacrificed: activities in the fields of science, engineering, architecture, law, art (with the exception of landscape and abstract paintings, in eco-oilpaint of course), trade, politics and so on would be suspect even if permitted. Sex would be illegal except during fertility festivals – after all, oestrogen is harmful to the environment, latex is made by hurting trees, and natural ways of preventing pregnancy are not secure enough to prevent overpopulation. The subjects open for public debate, and thereby the possibility to develop a plan of life and considered judgements and preferences in freedom, would be limited even without indoctrination. The agenda for demo-cratic decision-making would be reduced, in the worst case to that of the good old People's Democracies. At any rate, the freedom to transform society away from the ideal would be gone. Relations with other communities would be diminished – if not limited by a prescription to prevent unnecessary pollution caused by un-necessary travel, then simply to avoid contact with and import of all too unorthodox ideas. If worse were to come to the worst and the whole traditional apparatus of totalitarianism would still be insufficient to warrant that social and cultural condi-tions create the kind of individuals who fit the ideal society, there would be one last resort: genetic engineering. Unfortunately, because of its being an abomination in the eyes of Mother Nature, that technique would no longer be available.

Now *of course* this is all rhetoric, some of it very cheap, and *of course* it is a worst-case scenario and *of course* an ecologically balanced society can be a pleasant place to live in, at least for some of us. Yet if the worst possible futures of our present society (the reports of the Club of Rome, the impending ecological catas-trophe, a global Manhattan) are held against liberal democracy, it is only fair that we compare these to the worst possible futures under alternative circumstances. Moreover, if we feel that a green interpretation of liberalism like the one pres-ented here, showing that compromises between modern liberalism and environ-mentalism are possible, is not enough to serve as a basis of social critique and reform, if we feel that a 'greened' liberal democracy is insufficient to warrant sustainability, if we feel that with or without a free market (again, compromises were possible) there is still danger because the presence of ecology-conscious citizens is a necessary condition for the ecologically balanced society – if we feel all this, then there are only two alternatives left. Either we choose to try to get a grip on the minds and opinions of humans, taking the first steps on a road to serfdom, or we take up the challenge. It takes courage to accept that there is no reasonable alternative to liberal democracy.

9.5 Concluding remarks

Popular opinion has it that our world is facing if not an environmental crisis then at least environmental problems of immense magnitude and complexity. True or not, what counts is that this is a widely shared belief, a belief which, together with the political theories and practices it inspires, challenges the credibility of liberalism and the legitimacy of liberal democracy. So far, a coherent liberal theory of the sustainable society has been lacking; this book should come some way towards opening the debate on the subject among liberals and other supporters of liberal democracy. It has given evidence to the fact that green liberalism can be both a possible and a powerful political theory. The most popular interpretation of sustain-ability actually appears to be far less green than liberalism itself. Green liberalism

225

is also far less conservative and restrictive than so-called greenbashers (e.g. Adler 1973; O'Rourke 1994; Treanor 1996) argued environmentalism must be. It is an open-ended theory.

The good news, then, is that neither the philosophy of liberalism nor its political institutions are necessary conditions for the existence or persistance of an environmental crisis; in our pluralistic world, green liberalism is in fact a necessary condition for its solution. The bad news is that a greener liberalism is still insufficient to safeguard sustainability. I have argued that the blame for this cannot be put on economic liberalism alone. The institutions of the free market are factors that at best sustain and at worst promote environmentally imprudent preferences and behaviour. I have also argued that there is little hope for a remedy in injecting society with more democracy of a higher quality. The root of environmental problems in liberal democracies is in the individuals' economy of preferences, their genesis and justification. Engaging in a dialogue between the preferences of individuals and environmental side constraints is a necessary condition for the creation of a sustainable society – *any* society, not just liberal democracies. Political or economic institutions can only facilitate this dialectic process in the hearts and minds of individuals; they cannot do it in their place. Ultimately, the development of sustainable preferences, of a sustainable attitude and character, is the responsibility of each and every single individual.

Apart from this not too optimistic conclusion, this book may also fall short of expectation in other respects. I have deliberately ignored the spiritual side of ecologism, being unable to take the spiritual side of anything seriously. I have also ignored ecofeminism, I have refrained from analysing, let alone criticizing, the green critique of Western culture (its materialism, hedonism, arrogance, alienation from nature), I have dodged the question of what exactly distinguishes needs from wants, I have failed to set out a precise programme of action for a liberal democratic environmental policy or an agenda for a philosopher's debate, I have probably insulted advocates of the intrinsic value of nature, admirers of native cultures and peoples, vegetarians, vegans, animal rights activists and many other people of presumably good will. There is only so much text an author can put in a book, and only a limited number of private assumptions and prejudices he can question simultaneously.

Tradition prescribes that we should end a book with a remark about the necessity of further research or with a positive message of hope and something to the effect that we should all be good people and nice to one another. There is certainly a need for further research. Given that liberalism cannot prescribe the one ideal shape of a sustainable society, we need to know in more detail which different conceptions of sustainability are compatible with exactly which types of theory of social justice. Explorations in this field are already being conducted (Dobson forthcoming). Given that non-market solutions may simply transform a free market into a black market, we need to continue the debate on how the free market can facilitate a greener economy, yet rid it of its often all too ideological climate. And given the crucial role attributed to the individual's responsibility for their preferences, we need to know more about the virtues that are required to permit sustainability. Again, this is a terrain in which new contributions are to be expected shortly (Barry forthcoming).

As to the bit about messages of hope and calls for the Christmas spirit, I hope the reader will not be too disappointed if I end in a far more modest mood than the ambitious enterprise I have undertaken in this book might lead us to expect. Liberalism rests on the admission of our own fallibility and on the willingness to make a distinction between personal opinions on the good and the right on the one hand, and on the other ideas about the restrictions and commandments that a liberal democratic regime can legitimately impose on its subjects. In writing this book, I sometimes had to accept conclusions I did not like, precisely because of this distinction. If given a choice between a global Manhattan and a global Yellowstone Park, I would definitely prefer the former. Yet from the public point of view, this is the least desirable of the two: it reduces human freedom far more than the latter – which could still develop into an infinite set of privately preferable worlds including the global Manhattan. As I deeply dislike rabbit fodder, I was pleased to find that being carnivorous is not a mortal sin, but I feel quite uncomfortable about the discovery that contingent circumstances like our current treatment of animals still make us an accessory to essentially evil practices. The friction between public and private reached a high when I was forced to embrace a rigidly liberal position on the crucial issue of population policies. I would much have preferred to argue that overpopulation will remain a problem as long as peer pressure by family and friends reproduces the fable that having children is socially desirable, normal or expected, that it should be attributed to an inborn desire or, worst of all, that it is the fulfilment of a life, a necessary condition for being a complete person. On each of these points and many others, I was obliged to restrain myself. Questions of individual morality exist to be debated, not to be resolved authoritatively. If there is one respect in which I could inspire the reader, I hope it is this one.

Notes

1 The *locus classicus* for this approach, at least in political theory, is Robert Nozick's *Anarchy, State and Utopia* (Nozick 1974). It is also, in one form or other, the unquestioned basis of both the 'weak' and the 'strong' economic views on sustainable development: in the former, all resources are understood to be fungible; in the latter, natural and artificial resources are internally but not mutually intersubstitutable (cf. Jacobs 1995; Dobson 1996; Holland 1996).

2 This is one of the arguments used to defend the view that political issues should be decided by representatives rather than by plebiscite.

3 Please note that there is a difference between *a* common good and *the* common good. Liberal democracy is predicated on the recognition of value pluralism, of a plurality of equally worthy views on the common good; hence, it allows for a distinction between a purely subjective, even egoistic will and an objective will as understood from the perspective of *any* of these theories of the good, but not for a distinction between the subjective will and *the* common good. Policies and rules defended in terms of the latter concept should be seen as either remnants of a preliberal attitude or as attempts of majorities to monopolize morality.

4 For more on this, compare e.g. the debate between Rawls (1972, 1982), Sen (1973, 1980, 1990), Dworkin (1981) and others on the interpretation of equality as a criterion for justice.

5 This violates condition ℭ (Section 1.3) but it is the example that counts.

6 See Day 1983: 17; Flathman 1983: 1, 23; cf. Steiner 1983: 83. Classic philosophical conceptions of freedom like Spinoza's, Rousseau's and Hegel's, where freedom is understood (roughly) as the recognition of necessity, are special cases of this view on positive freedom.

7 Cf. Sen (1988, 1990, 1991, 1993a, 1993b). Isaiah Berlin (1969) seems unable to choose between both interpretations of positive freedom. For a further analysis of the two see Flathman 1987.

8 Other examples can be found in e.g. Wenz (1988), Dobson (1990), Bosselmann (1992), Wissenburg (1993) and Freeden (1995).

9 For further introductory reading on types of moral theory see e.g. Timmons (1987).

10 There is a direct link here between modern Anglo-Saxon liberalism and American pragmatism: both question the objectivity of subjective theories of truth. Pragmatism, unlike liberal political philosophy, links the observation that mind and its products develop in dialogue with their environment(s) to a sceptical attitude with regard to given preferences. It thus has a far less disinterested perspective on (the value of) nature than liberalism (cf. Parker 1996; Rosenthal 1996).

11 The alternative is a Levinas-like approach to moral relevancy: the recognition of another subject as morally relevant because the other is genuinely Other (Cooper 1995).

12 For more objections to this version of independent value, see e.g. Regan (1992).

13 Compare for this distinction Regan (1992).

14 In the unlikely event that we can believe in the existence of a creature x that has no life as defined here but still has a plan of 'life, Jim, but not as we know it', our definition of life excludes x from being considered a subject on technical and not substantial grounds, i.e. because x and we cannot interact or be conscious of one another's existence.

15 There is nothing odd about attributing rights to animals (or dead objects), even though animals cannot be conscious of their rights, let alone effectuate them; after all, we do the same in everyday language with the comatose or pseudo-comatose (our stoic Buddhist). It does however violate condition $\mathfrak{C}$ (Section 1.3). In formal terms, it would be more correct to represent the rights of animals as limits to the rights of humans: a monkey's right to life would then become either the absence of a permission to kill it or the presence of a prohibition on killing, as part of the set of rights of all humans.

16 To be exact, until Wissenburg (forthcoming). For classic defences of these elements of liberal democracy see e.g. Rawls (1972, 1993), Barry (1989, 1995a), Ackerman (1980), Nozick (1974), Galston (1980).

17 I use the term property in a technical sense: a property right is the sum of (1) a set of user rights indicating what we are permitted to do with an object, including or excluding rights to let or lend, and (2) a right to irrevocably transfer these user rights, i.e. to sell, give or bequeath.

18 For a detailed critique of Nozick's theory of entitlement see O'Neill (1981).

19 An act that must necessarily be performed to be able to perform another act is, by definition (cf. Section 1.3) not a separate 'basic act'.

20 In technical terms, I assume that conditional rights are explicitly attributed; the liberty to destroy or not destroy something is therefore represented as MayDo$(x$ & $\neg x)$. Since the option of doing x has been excluded, what remains is a permission to do $\neg x$ only, i.e. a duty. If rights were not explicitly attributed, we would have a freedom to do x or $\neg x$: $\neg$MayDo$\neg(x$ & $\neg x)$. Excluding the permission to do x from this right does *not* result in a duty to do $\neg x$.

21 Two, in fact. Another, more technical problem concerns choices between two sets, say $\{a, b, c\}$ and $\{b, c, d\}$, each of which consists of three acts that are together (i.e. as a set) necessary to perform a further legitimate act e. An example could be the choice between {driving to the park, walking to a field, getting undressed} and {walking to a field, getting undressed, taking the bus to the park} in order to do e, get a natural tan all over. Let us assume, for the sake of argument, that e is a basic need covered by an unconditional right. Both sets involve an act that damages nature (the use of petrol), but one that cannot be avoided since e is a basic need. A choice

between these sets cannot be made on the basis of the number of basic acts that would be excluded (the amount of freedom that we would be deprived of) by rejecting one set since both sets are of equal size. I would suggest, then, that a liberal decision rule in cases like these should look at the effects of performing each of these sets on the number of options or basic acts that remain open to others.

22 Andrew Dobson, personal communication.

23 Robert Goodin, personal communication.

24 As criticized by a.o. Brian Barry (Barry 1977: 268, 271) and Jan Narveson (1978: 56).

25 Another way to represent this problem is to point to the fact that we enter the real world, finding goods and institutions bequeathed to us by our predecessors without the express condition that we do the same for those who come after us. Why then, the critic would ask, should we feel obliged to do the same for our own descendants?

26 All this is not to say that an appeal to self-interest and mutual advantage offer the only way to give rise to principles of justice between generations. Justice is perhaps 'a good thing quite apart from its general long-run tendency to be in everyone's interest' (Barry 1978: 217) and there may be other arguments leading to such principles – but all I want to show here is that self-interest is a *sufficient* condition for their existence.

27 However, the fact that moral principles may be difficult to satisfy under harsh empirical conditions does not *invalidate* them (cf. Barry 1978: 212).

28 *Our Common Future*, World Commission on Environment and Development, quoted in Nelissen, Van der Straaten & Klinkers, 1997: 282.

29 This includes the option of placing the operator $\neg$ somewhere in what we might suppose to be an initial full liberty to procreate [CanDop, & CanDo$\neg p$,] & [MayDop, & MayDo$\neg p$.].

30 Fortunately, because these reasons are a mystery to me. There may not even be reasons, just a lack of self-control, or, worse, an uncontrolable urge to be a parent (cf. McCleary 1945).

31 The 'full plan of life' refers to that part of an individual's plan of life that does not depend on *particular* circumstances for its success. For instance, it does not take a technologically advanced society, a wheelchair or £1 000 000 to be a benevolent person, benevolence being part of the full plan of life (cf. Section 5.1). For more details, see Wissenburg (forthcoming).

32 Hypothetically speaking, at least in this section: any reason why mankind cannot survive must needs be a reason why at least some individuals cannot survive. I make the distinction between individuals and mankind because a resource that is not necessary for present human beings may be necessary for non-existent future generations or for 'other subjects' (animals), whose survival may be necessary for mankind.

33 Like animals migrating to an area where the existing species had and the immigrant species has no natural enemies; or like the meteor that is supposed to have at least contributed to the extinction of the dinosaurs.

34 It is probably true that 'the beneficiary pays' sounds less attractive than 'the polluter pays'. The popularity of the latter principle, despite its blatant disregard for questions of responsibility, is perhaps best explained by the fact that it alliterates.

35 At the collective level (see Chapter 6) there is nothing a liberal democratic polity can do – not even express disapproval.

36 Note, by the way, that the precautionary principle is as unsuccessful in dealing with pure (i.e. non-quantifiable) uncertainty as any other type of cost-benefit analysis must be.

37 This, in a way, is what ecological modernization (Weale 1992) is about, most of the time.

38 Both for financial and technical reasons and because developments in space technology may not keep pace with the development of environmental problems.

39 Like selling a limited amount of pollution rights (vouchers) to automobile owners.

40 Soylent Green is the name of a science fiction movie and also of the company that produces the synthetic food on which an overpopulated world of the future would have to live. It is made by recycling dead humans.

41 For a refreshing discussion of experiences with participatory democracy in communities and firms, supporting the points made here, see Carter (1996).

References

Achterberg, W. 1993. Can liberal democracy survive the environmental crisis? Sustainability, liberal neutrality and overlapping consensus. See Dobson & Lucardie (1993), 81–101.

Achterhuis, H. 1994. The lie of sustainability. See Zweers & Boersema (1994), 198–203.

Ackerman, B. A. 1980. *Social justice in the liberal state.* New Haven: Yale University Press.

Ackerman, B. A. 1983. What is neutral about neutrality? *Ethics* **93**, 372–90.

Adler, C. A. 1973. *Ecological fantasies.* New York: Green Eagle Press.

Agar, N. 1995. Valuing species and valuing individuals. *Environmental Ethics* **17**, 397–415.

Anderson, T. & D. Leal 1991. *Free market environmentalism.* Boulder: Westview Press.

Anderson, T. & D. Leal 1994. Freedom and the environment: a reply to critics. *Critical Review* **8**, 461–5.

Aristotle 1959. *Ethica Nicomachea*, (ed.) I. Bywater (1894). Oxford: Clarendon Press.

Aristotle 1980. *Nicomachean ethics*, (ed.) D. Ross. Oxford: Oxford University Press.

Arrow, K. J. 1995. A note on freedom and flexibility. In *Choice, welfare and development: a Festschrift in honour of Amartya K. Sen*, K. Basu, P. Pattanaik & K. Suzuruma (eds), 7–16. Oxford: Clarendon Press.

Attfield, R. 1988. Population policies and the value of people. See Hudson & Peden (1988), 191–202.

Bahro, R. 1980. *Elemente einer neuen Politik.* Berlin: Olle & Wolter.

Barry, B. M. 1965. *Political argument.* London: Routledge & Kegan Paul.

Barry, B. M. 1977. Justice between generations. In *Law, morality and society: essays in honour of H.L.A. Hart*, P. M. S. Hacker & J. Raz (eds), 268–84. Oxford: Clarendon Press.

Barry, B. M. 1978. Circumstances of justice and future generations. In *Obligations to future generations*, B. M. Barry & R. I. Sikora (eds), 204–48. Philadelphia: Temple University Press.

Barry, B. M. 1989. *A treatise on social justice I: theories of justice.* London: Harvester Wheatsheaf.

Barry, B. M. 1993. Intergenerational justice in energy policy. In *Justice*, M. Fisk (ed.), 223–37. Atlantic Highlands: Humanities Press.

Barry, B. M. 1995a. *A treatise on social justice II: justice as impartiality.* Oxford: Clarendon Press.

Barry, B. M. 1995b. Spherical justice and global injustice. In *Pluralism, justice and equality*, D. Miller & M. Walzer (eds), 67–80. Oxford: Oxford University Press.

REFERENCES

Barry, J. forthcoming. *Rethinking green politics: nature, virtue and progress.*

Barry, B. M. & R. E. Goodin (eds) 1992. *Free movement: ethical issues in the transnational migration of people and of money.* New York: Harvester Wheatsheaf.

Bayles, M. D. 1976a. Introduction: moral philosophy and population issues. See Bayles (1976c), xi-xvi.

Bayles, M. D. 1976b. Limits to a right to procreate. See Bayles (1976c), 41–55.

Bayles, M. D. (ed.) 1976c. *Ethics and population.* Cambridge, Mass.: Schenkman.

Beckerman, W. 1994. 'Sustainable development': is it a useful concept? *Environmental Values* **3**, 191–209.

Beckerman, W. 1995. How would you like your 'sustainability', Sir? Weak or strong? A reply to my critics, *Environmental Values* **4**, 168–79.

Beckerman, W. 1996. Sustainable development and inter-generational egalitarianism. Paper presented at the Environmental Sustainability and Social Justice Seminars, Keele, England.

Benton, T. 1993. *Natural relations: ecology, animal rights and social justice.* London: Verso.

Berlin, I. 1969. *Four essays on liberty.* Oxford: Oxford University Press.

Bosselmann, K. 1992. *Im Namen der Natur: der Weg zur ökologischen Rechtsstaat.* Darmstadt: Wissenschaftliche Buchgesellschaft.

Brandt, R. B. 1979. *A theory of the good and the right.* Oxford: Clarendon Press.

Buchanan, J. M. 1975. *The limits of liberty: between anarchy and Leviathan.* Chicago: Chicago University Press.

Callahan, D. 1976. Ethics and population limitation. See Bayles (1976c), 19–40.

Callenbach, E. 1975. *Ecotopia: the notebooks and reports of William Weston.* New York: Bantam Books.

Carens, J. H. 1992. Migration and morality: a liberal egalitarian perspective. See Barry & Goodin (1992), 25–47.

Carter, I. 1995a. The independent value of freedom. *Ethics* **105**, 819–45.

Carter, I. 1995b. Interpersonal comparisons of freedom. *Economics and Philosophy* **11**, 1–23.

Carter, I. 1997. The concept of freedom in the work of Amartya Sen: an alternative analysis consistent with freedom's independent value. *Notizie di Politeia* **12**, 7–22.

Carter, N. 1996. Worker co-operatives and green political theory. See Doherty & De Geus (1996), 56–75.

Cohen, J. 1997. Deliberation and democratic legitimacy. In *Contemporary political philosophy: an anthology*, R. E. Goodin & P. Pettit (eds), 143–55. Oxford: Basil Blackwell.

Common, M. S. 1996. Beckerman and his critics on strong and weak sustainability: confusing concepts and conditions. *Environmental Values* **5**, 83–8.

Cooper, D. E. 1995. Other species and moral reason. In *Just environments: intergenerational, international and interspecies issues*, D. E. Cooper & J. A. Palmer (eds), 137–48. London: Routledge.

Dahl, R. A. 1956. *A preface to democratic theory.* Chicago: Chicago University Press.

Daly, H. E. 1995. On Wilfred Beckerman's critique of sustainable development. *Environmental Values* **4**, 49–55.

Dan-Cohen, M. 1992. Conceptions of choice and conceptions of autonomy. *Ethics* **102**, 221–43.

Day, J. P. 1970. On liberty and the real will. *Philosophy* **45**, 177–92.

Day, J. P. 1983. Individual liberty. In *Of liberty*, A. Phillips Griffiths (ed.), 17–29. Cambridge: Cambridge University Press.

De Geus, M. 1993. *Politiek, milieu en vrijheid.* Utrecht: Jan van Arkel.

REFERENCES

De Geus, M. 1996. *Ecologische utopieën: ecotopia's en het milieudebat*. Utrecht: Jan van Arkel.

De Pizan, C. 1994. *The book of the body politic*. Cambridge: Cambridge University Press.

De-Shalit, A. 1996. A return to the scene of the crime: sustainability and intergenerational justice. Paper presented at the Sustainability and Social Justice Seminars, Keele, England.

De Tocqueville, A. 1951. *De la démocratie en Amerique*. Paris: Gallimard.

Devall, B. & G. Sessions 1984. The development of nature resources and the integrity of nature. *Environmental Ethics* **4**, 293–322.

Devall, B. & G. Sessions 1985. *Deep ecology: living as if nature mattered*. Salt Lake City: Gibbs M. Smith.

Dobson, A. N. H. 1990. *Green political thought: an introduction*. London: Unwin Hyman.

Dobson, A. N. H. 1995. *Green political thought: an introduction*, 2nd edn. London: Routledge.

Dobson, A. N. H. 1996. Sustainabilities: an analysis and a typology. *Environmental Politics* **3**, 401–28.

Dobson, A. N. H. forthcoming. *Justice and the environment: conceptions of environmental sustainability and theories of distributive justice*.

Dobson, A. N. H. & P. Lucardie (eds) 1993. *The political theory of nature*. London: Routledge.

Doherty, B. & M. De Geus (eds) 1996. *Democracy and green political thought: sustainability, rights and citizenship*. London: Routledge.

Dworkin, R. M. 1978. *Taking rights seriously*, 2nd edn. London: Duckworth.

Dworkin, R. M. 1981. What is equality? part 2: equality of resources. *Philosophy and Public Affairs* **4**, 283–345.

Dworkin, R. M. 1985. *A matter of principle*. Cambridge, Mass.: Harvard University Press.

Eckersley, R. 1996. Greening liberal democracy: the rights discourse revisited. See Doherty & De Geus (1996), 212–36.

Eden, S. 1996. *Environmental issues and business: implications of a changing agenda*. Chichester: John Wiley.

Elliot, R. 1992. Intrinsic value, environmental obligation and naturalness. *Monist* **75**, 138–60.

Ewald, F. 1986. *L'état providence*. Paris: Grasset & Fasquelle.

Fishkin, J. S. 1995. *The voice of the people: public opinion and democracy*. New Haven: Yale University Press.

Flathman, R. E. 1983. *The philosophy and politics of freedom*. Chicago: University of Chicago Press.

Francis, J. 1996. Nature conservation and the precautionary principle. *Environmental Values* **5**, 257–64.

Freeden, M. 1995. Green ideology: concepts and structures. Oxford: OCEES research paper No. 5.

Galston, W. A. 1980. *Justice and the human good*. Chicago: Chicago University Press.

Galston, W. A. 1991. *Liberal purposes: goods, virtues, and diversity in the liberal state*. Cambridge: Cambridge University Press.

Gauthier, D. 1985. Justice as social choice. In *Morality, reason and truth: new essays on the foundations of ethics*, D. Copp & D. Zimmerman (eds), 251–69. Totowa, N.J.: Rowman & Allanheld.

Gauthier, D. 1986. *Morals by agreement*. Oxford: Clarendon Press.

Gold, M. 1995. *Animal rights: extending the circle of compassion*. Oxford: Jon Carpenter.

Goodin, R. E. 1992. *Green political theory*. Cambridge: Polity Press.

Goodwin, B. 1992. *Justice by lottery*. Harlow: Harvester Wheatsheaf.

Gore, A. 1992. *Earth in the balance: ecology and the human spirit*. Boston: Houghton Mifflin.

Graham, E. & S. Graham 1992. The crisis of population growth. *International Journal of Moral and Social Studies* **7**, 205–36.

Gravel, N. 1994. Can a ranking of opportunity sets attach an intrinsic importance to freedom of choice? *American Economic Review* **84**, 454–80.

Gray, J. 1980. On negative and positive liberty. *Political Studies* **28**, 507–26.

Hardin, G. 1976. The tragedy of the commons. See Bayles (1976c), 3–18.

Hargrove, E. C. 1992. Weak anthropocentric intrinsic value. *Monist* **75**, 183–207.

Harlow, E. 1992. The human face of nature: environmental values and the limits of anthropocentrism. *Environmental Ethics* **14**, 27–42.

Harrison, P. 1997. The third revolution: environment, population and a sustainable world. Excerpt, see Nelissen, Van der Straaten & Klinkers (1997), 364–70.

Hartmann, W. K. 1984. Space exploration and environmental issues. *Environmental Ethics* **6**, 227–39.

Hauerwas, S. 1974. The moral limits of population. *Thought* **49**, 237–49.

Hayek, F. A. 1976. *The road to serfdom* (1944). London: Routledge & Kegan Paul.

Hessley, R. 1981. Should government regulate procreation: a third view. *Environmental Ethics* **3**, 49–53.

Hilhorst, M. 1994. The ethical assessment of new technologies: some methodological considerations. See Zweers & Boersema (1994), 262–71.

Holland, A. 1994. Natural capital. In *Philosophy and the natural environment*, R. Attfield & A. Belsey (eds), 169–82. Cambridge: Cambridge University Press.

Holland, A. 1996. Sustainability: or, why strong sustainability is weak and absurdly strong sustainability is not absurd. Paper presented at the Sustainability and Social Justice Seminars, Keele, England.

Holmes, S. 1993. *The anatomy of antiliberalism*. Cambridge, Mass.: Harvard University Press.

Hudson, Y. & C. Peden (eds) 1988. *Philosophical essays on the ideas of a good society*. Lewiston: Edwin Mellen Press.

Inglehart, R. 1977. *The silent revolution: changing values and political styles among Western publics*. Princeton: Princeton University Press.

Irvine, S. 1991. An overcrowded continent. In *Green Light on Europe*, Sara Parkin (ed.), 266–77. London: Heretic Books.

Jacobs, F. 1994. Can liberal democracy help us to survive the environmental crisis? See Zweers & Boersema (1994), 158–63.

Jacobs, M. 1995. Reflections on the discourse and politics of sustainable development, part I: faultlines of contestation and the radical model. Paper presented at the Sustainability and Social Justice Seminars, Keele, England.

Kanger, H. & S. Kanger 1966. Rights and parliamentarism. *Theoria* **2**, 85–115.

Kant, I. 1928. *Kritik der praktischen Vernunft*, (1788). Wiesbaden: VMA-Verlag.

Kant, I. 1974. *Kritik der praktischen Vernunft und Grundlegung zur Metaphysik der Sitten*. Frankfurt/Main: Suhrkamp.

Kymlicka, W. 1989. *Liberalism, community, and culture*. Oxford: Oxford University Press.

Kymlicka, W. 1990. *Contemporary political philosophy*. Oxford: Oxford University Press.

Kymlicka, W. 1995. *Multicultural citizenship: a liberal theory of minority rights*. Oxford: Clarendon Press.

Lafferty, M. & J. Meadowcroft (eds) 1996. *Democracy and the environment: problems and prospects*. Cheltenham: Edward Elgar.

Larrère, C. 1997. *Les philosophies de l'environnement*. Paris: Presses Universitaires de Paris.

Lauber, V. 1978. Ecology, politics and liberal democracy. *Government and Opposition* **14**, 199–217.

Lee, D. 1979. Some ethical decision criteria with regard to procreation. *Environmental Ethics* **1**, 65–9.

Lee, K. 1993. Instrumentalism and the last person argument. *Environmental Ethics* **4**, 333–44.

Lee, K. 1996. Homo faber: the ontological category of modernity. Unpublished paper.

Lindahl, L. 1977. *Position and change: a study in law and logic*. Dordrecht: Reidel.

Locke, J. 1973. *Two treatises of government*. London: Everyman.

Lovelock, J. 1979. *Gaia: a new look at life on Earth*. Oxford: Oxford University Press.

Mathews, F. 1991. *The ecological self*. London: Routledge.

Mathews, F. (ed.) 1996. *Ecology and democracy*. London: Frank Cass.

McCleary, G. G. 1945. Why does *Homo Sapiens* reproduce? *Hibbert Journal* **46**, 239–42.

Mill, J. S. 1965. *Principles of political economy: with some of their applications to social theory*. London: Routledge & Kegan Paul.

Milsted, D. 1990. *The green bluffer's guide*. Horsham: Ravette Books.

Moore, G. E. 1993. *Principa ethica*, revised edn. Cambridge: Cambridge University Press.

Næss, A. 1989. *Ecology, community and lifestyle: outline of an ecosophy*. Cambridge: Cambridge University Press.

Nagel, T. 1986. *The view from nowhere*. Oxford: Oxford University Press

Nagel, T. 1991. *Equality and partiality*. Oxford: Oxford University Press.

Narveson, J. 1973. Moral problems of population. *Monist* **57**, 62–86.

Narveson, J. 1976. Moral problems of population. See Bayles (1976c), 59–80.

Narveson, J. 1978. Future people and us. In *Obligations to future generations*, B. M. Barry & R. I. Sikora (eds), 38–60. Philadelphia: Temple University Press.

Narveson, J. 1995. The case for free market environmentalism. *Journal of Agricultural and Environmental Ethics* **8**, 145–56.

Nas, M. 1995. Green, greener, greenest. In *The impact of values*, J. van Deth & E. Scarbrough (eds), 275–300. Oxford: Oxford University Press.

Neefjes, K. 1996. Ecological degradation: a cause for conflict, a concern for survival. Paper presented at the Sustainability and Social Justice Seminar Series, Keele, England.

Nelissen, N., J. Van der Straaten & L. Klinkers (eds) 1997. *Classics in environmental studies: an overview of classic texts in environmental studies*. Utrecht: International Books.

Norton, B. 1982a. Environmental ethics and nonhuman rights. *Environmental Ethics* **41**, 17–36.

Norton, B. 1982b. Environmental ethics and the rights of future generations. *Environmental Ethics* **4**, 317–37.

Norton, B. 1994. Economists' preferences and the preferences of economists. *Environmental Values* **3**, 311–32.

Norton, B. 1996. Ecology and freedom: towards a theory of sustainable opportunities. Paper presented at the Sustainability and Social Justice Seminars, Keele, England.

Nozick, R. 1974. *Anarchy, state, and utopia*. New York: Basic Books.

O'Neill, J. 1992. The varieties of intrinsic value. *Monist* **75**, 119–37.

O'Neill, O. 1979. Begetting, bearing, and rearing. In *Having Children*, O. O'Neill (ed.), 25–38. Oxford: Oxford University Press.

REFERENCES

O'Neill, O. 1981. Nozick's entitlements. In *Reading Nozick: essays on anarchy, state, and utopia*, P. Jeffrey (ed.), 305–22. Oxford: Basil Blackwell.

Opschoor, H. 1994. Market forces as causes of environmental degradation. See Zweers & Boersema (1994), 175–97.

O'Riordan, T. & A. Jordan 1995. The precautionary principle in contemporary environmental politics. *Environmental Values* **4**, 191–212.

O'Rourke, P. J. 1994. *All the trouble in the world: the lighter side of famine, pestilence, destruction and death.* London: Picador.

Paine, T. 1989. *Political writings.* Cambridge: Cambridge University Press.

Parent, W. A. 1974. Some recent work on the concept of liberty. *American Philosophical Quarterly* **11**, 149–67.

Parfit, D. 1984. *Reasons and persons.* Oxford: Clarendon Press.

Parker, K. A. 1996. Pragmatism and environmental thought. In *Environmental Pragmatism*, A. Light & E. Katz (eds), 21–37. London: Routledge.

Pattanaik, P. & Y. Xu 1990. On ranking opportunity sets in terms of freedom of choice. *Recherches Economiques de Louvain* **56**, 383–90.

Pattanaik, P. & Y. Xu 1995. On preference and freedom. Unpublished paper.

Popper, K. R. 1960. *The poverty of historicism*, 2nd edn. London: Routledge & Kegan Paul.

Publius (Alexander Hamilton, James Madison, John Jay) 1962. *The federalist papers.* New York: New American Library.

Pufendorf, S. 1991. *On the duty of man and citizen according to natural law.* Cambridge: Cambridge University Press.

Puppe, C. 1994. An axiomatic approach to 'preferences for freedom of choice'. Unpublished paper.

Rawls, J. B. 1972. *A theory of justice.* Oxford: Oxford University Press.

Rawls, J. B. 1982. Social unity and primary goods. In *Utilitarianism and beyond*, A. K. Sen & B. Williams (eds), 159–85. Cambridge: Cambridge University Press.

Rawls, J. B. 1990. The island of the unborn. In *Time-travel for beginners and other stories*, J. Harrison (ed.), 28–30. Nottingham: Nottingham University Press.

Rawls, J. B. 1993. *Political liberalism.* New York: Columbia University Press.

Rawls, J. B. 1995. Reply to Habermas. *Journal of Philosophy* **92**, 132–80.

Raz, J. 1986. *The morality of freedom.* Oxford: Clarendon Press.

Regan, T. 1984. *The case for animal rights.* London: Routledge.

Regan, T. 1992. Does environmental ethics rest on a mistake? *Monist* **75**, 161–82.

Rosenthal, S. B. 1996. How pragmatism *is* an environmental ethic. In *Environmental Pragmatism*, A. Light (ed.), 38–49. London: Routledge.

Rothbard, M. N. 1973. *For a new liberty: the libertarian manifesto.* New York: Macmillan.

Routley, V. & R. Routley 1982. Nuclear power: some ethical and social dimensions. In *And justice for all*, T. Regan & D. VanDeVeer (eds), 116–38. Totowa: Rowman & Littlefield.

Samaras, T. 1995. *The truth about your height.* San Diego: Tecolote Publications.

Sandel, M. 1982. *Liberalism and the limits of justice.* Cambridge: Cambridge University Press.

Saward, M. 1996. Must democrats be environmentalists? See Doherty & De Geus (1996), 79–96.

Sen, A. K. 1970. *Collective choice and social welfare.* San Francisco: Holden-Day.

Sen, A. K. 1973. *On economic inequality.* Oxford: Clarendon Press.

REFERENCES

Sen, A. K. 1980. Equality of what? In *The Tanner Lectures on Human Values 1980*, S. M. McMurrin (ed.), 196–220. Salt Lake City: University of Utah Press.

Sen, A. K. 1988. Freedom of choice: concept and content. *European Economic Review* **32**, 269–94.

Sen, A. K. 1990. Welfare, freedom and social choice: a reply. *Recherches Economiques de Louvain* **56**, 451–85.

Sen, A. K. 1991. Welfare, preference and freedom. *Journal of Econometrics* **50**, 15–29.

Sen, A. K. 1993a. Markets and freedoms: achievements and limitations of the market mechanism in promoting individual freedoms. *Oxford Economic Papers* **45**, 519–41.

Sen, A. K. 1993b. Capability and well-being. In *The Quality of Life*, M. Nussbaum & A. K. Sen (eds), 30–53. Oxford: Clarendon Press.

Sen, A. K. 1995. Rationality and social choice. *American Economic Review* **85**, 1–24.

Singer, B. 1988. An extension of Rawls' theory of justice to environmental ethics. *Environmental Ethics* **3**, 217–31.

Singer, P. 1977. *Animal liberation: towards an end to man's inhumanity to animals*. London: Johnathan Cape.

Smith, T. 1995. The case against free market environmentalism. *Journal of Agricultural and Environmental Ethics* **8**, 126–44.

Spinoza, B. 1951. *Tractatus Theologico-politicus and Tractatus Politicus*. New York: Dover.

Spinoza, B. 1972. *Ethica Ordine Geometrico Demonstrata*. In *Opera*, 2nd edn, C. Gebhardt (ed.). Heidelberg: Carl Winters Universitätsbuchhandlung.

Steiner, H. 1983. How free: computing personal liberty. In *Of liberty*, A. Phillips Griffiths (ed.), 73–89. Cambridge: Cambridge University Press.

Steiner, H. 1992. Libertarianism and the transnational migration of people. See Barry & Goodin (1992), 87–94.

Stephens, P. 1996. Plural pluralisms: towards a more liberal political theory. Unpublished paper.

Sterba, J. P. 1996. Reconciliation reaffirmed: a reply to Steverson. *Environmental Values* **5**, 363–8.

Tabah, L. 1995. Population prospects with special reference to the environment. In *Just environments: intergenerational, international and interspecies issues*, D. E. Cooper & J. A. Palmer (eds), 72–88. London: Routledge.

Taylor, R. 1993. The environmental implications of liberalism. *Critical Review* **6**, 265–82.

Temkin, L. S. 1986. Inequality. *Philosophy and Public Administration* **2**, 99–121.

Temkin, L. S. 1993. *Inequality*. Oxford: Oxford University Press.

Tester, K. 1991. *Animals and society: the humanity of animal rights*. London: Routledge.

Thero, D. P. 1995. Rawls and environmental ethics: a critical examination of the literature. *Environmental Ethics* **17**, 93–106.

Timmons, M. 1987. Ethical foundationalism. *Ethics* **97**, 595–609.

Treanor, P. 1996. *Why sustainability is wrong*. http://web.inter.nl/users/Paul.Treanor/sustainability.html. Posted before 2–9–1996.

Unabomber (Freedom Club, allegedly T. J. Kaczynski) 1995. *Manifesto*. http://www.soci.niu.edu/~critcrim/uni/uni.txt. Posted 22–9–1995.

Van der Wal, K. 1994. Technology and the ecological crisis. See Zweers & Boersema (1994), 215–46.

Van Hees, M. V. B. P. M. 1995a. *Rights and decisions: formal models of law and liberalism*. Dordrecht: Kluwer Academic Publishers.

REFERENCES

Van Hees, M. V. B. P. M. 1995b. On the analysis of negative freedom. Unpublished paper.

Van Hees, M. V. B. P. M. & M. L. J. Wissenburg forthcoming. Opportunity and freedom. *Political Studies.*

Van Parijs, P. 1992. Citizenship exploitation, unequal exchange and the breakdown of popular sovereignty. See Barry & Goodin (1992), 155–65.

Vincent, A. 1993. The character of ecology. *Environmental Politics* 2, 248–76.

VROM, Ministerie van 1989. *Nationaal Milieubeleidsplan.* 's-Gravenhage: SDU.

Walter, E. 1988. Morality and population. See Hudson & Peden (1988), 203–16.

Walzer, M. 1983. *Spheres of justice.* New York: Basic Books.

Walzer, M. 1988. Interpretation and social criticism. In *The Tanner Lectures on Human Values 1985*, S. M. McMurrin (ed.), 1–80. Salt Lake City: University of Utah Press.

WCED (World Commission on Environment and Development) 1987. Our Common Future. Excerpt see Nelissen; Van der Straaten & Klinkers (1997), 275–84.

Weale, A. 1992. *The new politics of pollution.* Manchester: Manchester University Press.

Weenink, A. W. 1997. *De russische paradox: overvloed en onvermogen.* PhD Thesis, Enschede: Twente University Press.

Wells, D. T. 1995. *Environmental policy: a global perspective for the twenty-first century.* Upper Saddle River: Prentice Hall.

Wenz, P. S. 1988. *Environmental justice.* Albany: State University of New York Press.

Wissenburg, M. L. J. 1993. The idea of nature and the nature of distributive justice. See Dobson & Lucardie (1993), 3–20.

Wissenburg, M. L. J. 1996. European constitutional decision making: the case of green v. liberal democratic concerns. Nijmegen, The Netherlands: Nijmegen Political Science Reports nr. 36. To be published as Environmental protection in Europe. In *The Political Theory of European Constitutional Choice*, A. Weale & M. Nentwich (eds). London: Routledge 1998.

Wissenburg, M. L. J. 1997a. A taxonomy of green ideas. *Journal of Political Ideologies* 2, 29–50.

Wissenburg, M. L. J. 1997b. Past and future of green liberalism. In *Memory, history and critique: European identity at the millennium.* Proceedings of the 6th International ISSEI Conference at the University for Humanist Studies, Utrecht, The Netherlands, August 1996, F. Brinkhuis & S. Talmor (eds), CD-Rom. Harvard: MIT Press.

Wissenburg, M. L. J. 1997c. The possibility of green liberalism. In *Memory, history and critique: European identity at the millennium.* Proceedings of the 6th International ISSEI Conference at the University for Humanist Studies, Utrecht, The Netherlands, August 1996, F. Brinkhuis & S. Talmor (eds), CD-Rom. Harvard: MIT Press.

Wissenburg, M. L. J. forthcoming. *Imperfection and Impartiality: an outline of a liberal theory of social justice.* London: UCL Press.

Worster, D. 1985. *Nature's economy: a history of ecological ideas.* Cambridge: Cambridge University Press.

Zweers, W. & J. J. Boersema (eds) 1994. *Ecology, Technology and Culture: Essays in Environmental Philosophy.* Knapwell: White Horse Press.

Index

Printed and bound by CPI Group (UK) Ltd, Croydon, CR0 4YY

23/10/2024

01777665-0013